NUMBERS:
Arithmetic and Computation

Asok Kumar Mallik
Amit Kumar Das

Levant Books
India

First published 2022
by CRC Press
4 Park Square, Milton Park, Abingdon, Oxon, OX14 4RN

and by CRC Press
6000 Broken Sound Parkway NW, Suite 300, Boca Raton, FL 33487-2742

British Library Cataloguing-in-Publication Data
A catalogue record for this book is available from the British Library

Library of Congress Cataloging-in-Publication Data
A catalog record has been requested

ISBN: 9781032277592 (hbk)
ISBN: 9781003293941 (ebk)

DOI: 10.4324/9781003293941

Typeset in Knuth Computer Modern 10.5 pt
by Levant Books

LEVANT

Preface

For the first time within the same cover three topics, namely, numbers, arithmetic and computation are made available to the school and college students as well as non-expert adults who have interest in this area. This integration of the topics is deliberate as over the last few decades, computers, like mathematics for a few centuries, have become an integral part of basic education. Having both within the same cover is expected to help the reader better comprehend numbers and arithmetic, developed over the millennium. The reader will also appreciate how the recent proliferation of computers can help solving arithmetic problems using the age-old knowledge supported by modern technology.

So far as numbers are concerned, the discussion has been restricted to only integers and real numbers. Though the students at higher classes are exposed to complex numbers as well, complex numbers are not included keeping in mind the non-expert adults, most of whom are either not exposed to or forgot the complex numbers due to lack of use in daily life. But such adults are often interested in solving arithmetic problems involving integers and real numbers. To avoid scaring them away, complex numbers, like so many other types of numbers created by mathematicians over centuries, have been excluded. People interested in those numbers are referred to another book by the first author entitled 'The Story of Numbers'.

In chapter 1, we introduce different types of integers and real numbers, created by mathematicians of different ages and civilizations using different approaches and for different purposes. An unexpected and interesting relation between two types of numbers is shown to arouse overall interest of the readers regarding various types of numbers. Then we discuss some basic facts about integers and real numbers in chapters 2 and 3, respectively. These include definitions, representations, classifications, curious patterns of numbers and connections between various types of numbers. Also included are some important universal numbers, some basic facts and techniques required to solve problems. Several interesting stories regarding great mathematicians contributing to the field of numbers are also mentioned. We have referred to some apparently simple conjectures, where no counter-example has been found but are yet to be proved mathematically as true or false. We should clarify here that we have consciously included some inevitable repetitions of topics from the book 'The Story of Numbers', to make the current book self-contained.

In chapter 4, a large number of problems on integers and real numbers are given as exercises for practice. Most of these problems are not typical textbook problems. In a mathematics text, the concepts discussed in a chapter invariably guide the student for choosing the correct method to solve the end-of-the chapter exercise problems. The problems in chapter 4 are classified neither according to any particular concept, nor graded according to the level of difficulty. The level of difficulty ranges from easy to moderate, and moderate to difficult. It is hoped that these problems will sharpen the skill and mathematical thinking of the readers and also generate interest to continue with mathematics. Arithmetic solutions to all the problems are also included in chapter 5. However, the readers are cautioned that mathematics can be learnt only by doing and solving problems and not by mere reading the solutions. So, the readers are strongly encouraged to attempt these problems, even just by trying one learns more than by mere reading the solution without even making a serious effort to solve. Sometimes the reader may get a better solution than the one given in the book. Alternative solutions to certain problems are included to show that the solution procedure need not be unique, even if finally, one arrives at the same numerical answer. We have consciously included some good arithmetic problems from another book of the first author entitled 'Popular Problems and Puzzles in Mathematics'. These problems involve only integers and real numbers.

It is undeniable that over the last half a century, like mathematics over a few millennia, computers have become an inseparable part of human civilization. Initially it was primarily used for relieving humankind from the monotonous task of repetitive calculations. Gradually it has entered all spheres of human activities, much beyond arithmetic and mathematics. In chapter 6 of this book, we introduce some basic ideas about computers and programming languages used to communicate with the computers. Both the aspects are kept to the minimum as Computer Science and Engineering has become a different field independent of mathematics. The readers are expected to get a lot of fun while solving difficult arithmetic problems by using a computer; sometimes this may also be quicker and easier than the analytical method. After getting the answers by both arithmetic and computer, the readers may get better understanding, confidence and confirmation regarding the solutions so obtained for a new problem. Considering these reasons, the basic manipulations of different types of numbers possible through simple logic using powerful programming languages are discussed in this chapter. A reader is encouraged to either obtain a counter-example to disprove a conjecture, or widens the range of its validity. It must be mentioned there are problems which may well be beyond the analytical ability of some students, but they may be able to get the answer through computer. At the same time, some problems can only be solved by arithmetic and remain beyond the realm of programming.

In chapter 7, the knowledge gained in chapter 6 is used to solve some problems, given in chapter 4, by using a computer. Various types of problems are selected, some easier and quicker to solve by computer and some are shown not amenable to computer. It is further shown how some conjectures and theorems mentioned in chapters 2 and 3 can be verified by computers. The book ends with several appendices containing some details of topics included in previous chapters. This has been done so as not to obstruct the flow of the main ideas.

Several friends have helped us to complete this work. We gratefully acknowledge the effort and time spent by Prof. Manas Hira, Prof. Bhabatosh Chanda, Prof. Prateek Khastagir and Mr Siddhartha Sankar Chatterjee who read parts of the manuscript and gave their opinion so as to improve the quality. Special mention must be made of Dr Amit Kumar Sinhababu who helped with materials and suggestions to vastly improve chapters 1 to 5. Finally, we express our heartfelt gratitude to our friends Prof. Raminder Singh and Mr Rupak Biswas, who went through the entire manuscript and did an admirable job of editing the language. Only the authors are responsible for the mistakes that still remain. Last but not the least; we sincerely appreciate the patience and cooperation shown by Mr Abhijit Sen, who prepared the camera-ready version of the manuscript.

The authors will feel rewarded if the readers enjoy reading the book and solving the problems. We will be happy if the readers point out any mistake that eluded our scrutiny.

Kolkata
14 October 2021

Asok Kumar Mallik
asokiitk@gmail.com

Amit Kumar Das
amit.das.becs@gmail.com

v

Contents

Chapter 1

Introduction

We all start learning mathematics with counting objects as $1, 2, 3, \ldots$. This set of numbers, called the *natural numbers,* is denoted by the symbol $\mathbb{N}$. Famous German mathematician Leopold Kronecker (1823–91) once commented "God made the integers, man made the rest". What he meant is that some means of *knowing how many* existed long before mathematics was established by human beings. Primitive men knew how many cattle the family possessed. Curious experiments have established that even some animals and birds have the sense of small numbers to determine *how many.* In fact, it has been found that some of these creatures can be trained to count up to quite a few.

Mathematicians started with these natural numbers and extended the domain of numbers by creating abstract quantities, which exist only in the realm of human imagination. A great leap in the concept of numbers is credited to ancient Indian mathematicians, when they *created* a number *shunya,* universally called *zero* with the symbol 0. This is treated as a whole number representing the absence of an object. The set of natural numbers is also known as the *set of positive integers* written as $\mathbb{Z}^+$. The set of whole numbers is the union of 0 and $\mathbb{N}$ and is denoted by $\mathbb{W}$. As kids we are taught the symbol 0 along with the other counting numbers. Then we are taught addition of two elements of $\mathbb{W}$ and we get an element within this set. When teaching subtraction of a bigger number from a smaller one, we introduce negative integers like $-1, -2, -3, \ldots$. This set of negative integers is denoted by $\mathbb{Z}^-$. The set of integers $\mathbb{Z}$ includes 0, all the positive and negative integers. We will discuss these integers in some detail in Chapter 2.

After learning addition and subtraction, we are taught multiplication (or repeated additions). Multiplying two elements of $\mathbb{Z}$, we get an element of the same set. We know that the natural numbers continue forever, and in mathematics this unlimited, endless thing is called *infinity* and universally expressed by the symbol ∞. It must be pointed out that 0 and ∞ are symbols and abstract ideas and one has to be careful while using these in the midst of operations with counting

1

numbers. This is especially important when applying mathematics to solve real life problems. Mathematically the thickness of a line is zero, but any line drawn in real life will have some non-zero thickness.

The inverse of multiplication or division of an integer by another does not necessarily create another element within $\mathbb{Z}$. It is also important to note that division by 0 is not defined. The division of one integer by another (not zero) results in two integers, one we call the quotient and the other remainder. When the remainder is zero, we say the first one is divisible by the second. From this division process, a new type of number is created. This extension of numbers, called *rational numbers,* we will learn a little later when we learn fractions. Here we describe the process of division of one integer by another again by two integers, but not by the quotient and the remainder. Rather by a rational number (p/q), with $q \neq 0$, i.e. just write as the dividend by the divisor, both of which are integers.

It may be mentioned at this stage, that in school initially we learn Arithmetic, the rules of handling numbers with different operations, like addition, subtraction, multiplication, division, etc. Then we are taught other branches like Algebra and Geometry, which were developed in different ages and in different civilizations. An interesting feature of mathematics is that different branches originated in different ages and places may be connected at a later date. For example, you may have been exposed to the branch coordinate geometry, developed by the French philosopher and mathematician René Descartes (1596-1650) in the seventeenth century. This branch established connection between geometry, developed much earlier by the Greeks, with Algebra developed by the Arabs around the eighth century.

The Greek geometers were interested in mathematical descriptions and measurements of shapes and sizes. Questions like *How big is a field?, How long is a distance?*, etc. were more important than *how many?* Basically they were not concerned with discrete identical elements but something which is continuous. Geometry literally means measurement of the earth. The mathematics of the Greeks did not have any symbol for numbers. They had the idea of a point but not of zero.

It may not be out of place to mention that like the numbers, the ideal geometrical shapes are also imaginary or abstract concepts. No straight line or triangle can be drawn in real life; just as no one has seen a (2/3) donkey. As mentioned earlier, a geometrical line is supposed to have no thickness and we cannot draw a line of zero thickness. Whatever we draw must be at least a molecule thick. No wonder we always started our geometric propositions as *Let ABC be a triangle,* since a triangle can exist only in imagination, the figure drawn on a piece of paper or slate by us is not an object of geometry.

The whole of plane geometry is based on two figures, the straight line and the circle. Both these figures are defined by two points, say *A* and *B*. For drawing these two figures two instruments are available: (i) an unmarked straight edge for drawing a straight line joining A and B and, if necessary, extending the line beyond the segment AB on one or both sides; (ii) a compass to draw a circle with one of the points A (or B) as centre and passing through the other point B (or A). A point is located at the intersection of (i) two straight lines, (ii) two circles and (iii) a straight line with a circle. In the last two cases two points of intersection are normally generated, unless the two figures touch each other.

In geometry there is no measurement, as measurement is a physical action and hence *approximate* rather than an *exact* mental construction. Assigning unity to any arbitrary length of a straight line, whichever multiples or fractions of that length can be produced by geometrical construction the Greek mathematicians were ready to recognize that as a number. In this way they gave the status of number to all positive integers and rational numbers. A straight line of negative length is meaningless, so they did not have the concept of negative numbers. But it is not difficult to geometrically draw a line of length $\sqrt{x}$, if a line segment is assumed to represent the number x. So they believed in the operation of square root.

Pythagoreans believed that everything obtained by geometry can be expressed by positive integers (including their ratios), so they gave these the status of *God* and declared everything is a *number*. Someone clearly proved that the ratios of the lengths of the diagonal to the side of a square or a regular pentagon (both can be drawn geometrically) cannot be expressed as the ratio of two numbers! Now we know these values are $\sqrt{2}$ and $(\sqrt{5}-1)/2$, respectively. This came as a terrible shock to the Pythagoreans and they decided to conceal this fact from general public. It is said the traitor revealing it was thrown into the ocean as a punishment. All such numbers were given the name irrational numbers. The infinite number of integers and the rational fractions (p/q), together are called the *rational numbers*. If we represent these rational numbers geometrically by lengths on a line segment, the entire line segment is not filled up. We need all the irrational numbers to fill the gap and form a continuum called the *set of real numbers*, $\mathbb{R}$. Geometrically the real numbers are represented by a straight line of infinite length called the *real number line*. The details of real numbers are discussed in Chapter 3. In the second half of the nineteenth century the arithmetic rather than geometric concept of irrational numbers was developed. We may recall the famous statement of a great mathematician of this time, Richard Dedekind (1831–1916), who said *Numbers are free creations of the human mind; they serve as a means of apprehending more easily and more sharply the difference of things.*

Mathematics is about patterns and interconnections between various concepts. Before going into the details of integers and real numbers, we discuss one such example of curious connections. Towards this end, we first elaborate the way Greeks classified the positive integers. For them the geometrical shapes were important, accordingly they classified numbers as Triangular numbers or Square numbers. Triangular numbers 3, 6, 10, . . . , etc. are explained, respectively in Figs 1.1(a)–(c). These figures show how these many numbers of pebbles can be arranged in the shape of an equilateral triangle on the sand.

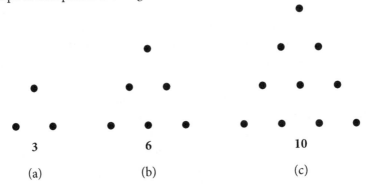

3   6   10

(a)   (b)   (c)

Fig. 1.1: Triangular numbers

Similarly, Fig. 1.2 explains the square numbers 4, 9, 16, . . . , where pebbles are arranged in the form of a square.

Now we will show how these diagrams can be used to obtain some modern day formulae, which most readers may have encountered, in a totally different way. It can be readily seen that the triangular numbers and square numbers can be written, respectively as

Triangular numbers

$$T = \frac{n(n+1)}{2} \tag{1.1}$$

where $n \in \mathbb{N}$.

Square numbers

$$S = m^2 \tag{1.2}$$

where $m \in \mathbb{N}$.

It may be noted that equations (1.1) and (1.2) both include the number 1 for $n = 1$ and $m = 1$, respectively as triangular and square numbers. But one pebble does not define any shape (or all shapes?). Consider the continuous (inclined) lines drawn in Figs 1.2(b) and (c) and count the dots above and below these lines to confirm the statement that the sum of two consecutive triangular numbers is a square number corresponding to the higher of the two. By modern algebraic notation we can reach the same conclusion as follows:

$$\frac{n(n+1)}{2} + \frac{(n+1)(n+2)}{2} = \frac{(n+1)(2n+2)}{2} = (n+1)^2 \tag{1.3}$$

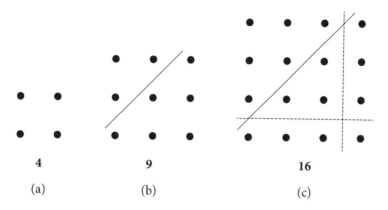

Fig. 1.2: Square numbers

By counting the four sets of dots separated by the dashed lines in Fig. 1.2(c) and the total numbers of dots it is readily seen that we get the well known formula of algebra

$$(m + 1)^2 = m^2 + m + m + 1 = m^2 + 2m + 1 \tag{1.4}$$

Now we will use these pictorial depictions of numbers to derive two well known formulae. First we obtain the sum of first consecutive n natural numbers from 1 to n. For this purpose, we count the dots along the diagonal line and lines parallel to the diagonal in Figs 1.2(a)–(c) and write

$$2^2 = 1 + 2 + 1$$
$$3^2 = 1 + 2 + 3 + 2 + 1$$
$$4^2 = 1 + 2 + 3 + 4 + 3 + 2 + 1$$

$$\vdots$$

$$n^2 = 1 + 2 + 3 + \cdots + (n - 1) + n + (n - 1) + \cdots + 3 + 2 + 1$$
or, $n^2 - n = 2[1 + 2 + 3 + \cdots + (n - 1)]$
or, $1 + 2 + 3 + \cdots + (n - 1) = (n^2 - n)/2 = n(n - 1)/2$
or, $1 + 2 + 3 + \cdots + n = n(n + 1)/2 \tag{1.5}$

Next we show that the sum of first m consecutive odd numbers is m^2. Here we recall how we get this result using our modern school algebra. We write the sum as

$$S = \sum_{l=1}^{m} (2l - 1) = 1 + 3 + 5 + \ldots (2m - 5) + (2m - 3) + (2m - 1)$$

Writing the above sum in reverse order

$$S = (2m - 1) + (2m - 3) + (2m - 5) + \cdots + 5 + 3 + 1$$

Now adding the above two expressions for S, we get

$$2S = 2m + 2m + 2m + \cdots + 2m + 2m + 2m = 2m^2 \text{ or, } S = m^2 \qquad (1.6)$$

This result is easily observed by counting the dots within the channels in Fig. 1.3, where square numbers are shown geometrically.

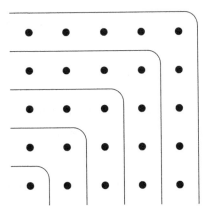

Fig. 1.3: Square numbers as the sum of two consecutive triangular numbers

Mathematics is a subject which delights in creating patterns and recognizing interconnections. Interconnection between various branches has already been mentioned, with reference to coordinate geometry. We will talk about curious patterns generated by numbers in later chapters. In the incredible, wonderful world of numbers, sometimes mysterious and totally unexpected connections are seen. We will just talk of one such connection now with reference to triangular numbers and square numbers defined above. It can be shown that certain numbers like 1 (with $n = 1$ in (1.1), $m = 1$ in (1.2)), 36($n = 8$, $m = 6$), 1225($n = 49$, $m = 35$), 41616($n = 288$, $m = 204$), etc. belong to both the groups of triangular and square numbers.

In 1733, one of the greatest mathematicians in history, Leonhard Euler (1707-83) raised the question for what values of m the square numbers m^2 are also triangular numbers and he also provided the answers written above. Basically he obtained the integer values of m and n which satisfy the following equation:

$$m^2 = \frac{n(n + 1)}{2} \qquad (1.7)$$

Some detailed discussion on how to solve this equation is included in Appendix A at the end of this book. This equation has infinite number of solutions. What is surprising is that all the infinite values of m can be obtained from the following continued fraction:

$$\cfrac{1}{6 - \cfrac{1}{6 - \cfrac{1}{6 - \ldots}}} \tag{1.8}$$

Consider the fractions obtained by terminating this infinite continued fraction at different steps (called the *rational convergent*), i.e. the approximate values of this continued fraction with higher and higher accuracy. These values are easily obtained as

$$\frac{1}{6}, \qquad \cfrac{1}{6 - \cfrac{1}{6}} = \frac{6}{35}, \qquad \cfrac{1}{6 - \cfrac{1}{6 - \cfrac{1}{6}}} = \frac{35}{204} \tag{1.9}$$

It is readily seen that the numerator and denominator of every rational convergent gives two consecutive values of m reported by Euler! Once the values of m's are obtained, the corresponding values of n can be easily obtained from (1.7). The continued fraction (solution) given by (1.8) was obtained by the great Indian genius Srinivasa Ramanujan (1887–1920). There is an interesting story associated with Ramnujan's solution as mentioned in his biography entitled *The Man who Knew Infinity* authored by Robert Kanigel.

During the year 1914 Ramanujan was in Cambridge. On a Sunday morning, Prasanta Chandra Mahalanobis (1893–1972 — Founder Director of the Indian Statistical Institute and a great Statistician) came to meet him with a puzzle published in the English newspaper *The Strand*. The puzzle stated that there are houses numbered serially $1, 2, 3, \ldots, n$ on one side of a street. The house numbered m is such that the sum of the house numbers on the left of that house is the same as the sum of the house numbers on the right side of that house. It is given that $50 < n < 500$, determine the number m. If we restate the puzzle mathematically, we may write that, it is given that

$$1 + 2 + 3 + \cdots + (m - 1) = (m + 1) + (m + 2) + \cdots + (n - 1) + n$$

$$\text{or,} \quad \frac{m(m - 1)}{2} = \frac{(n - m)(n + m + 1)}{2}$$

$$\text{or,} \quad m^2 = \frac{n(n + 1)}{2} \tag{1.10}$$

which is same as (1.7) and we have to find the integer values of m for which n is another integer. In other words, this puzzle is same as the question posed by Euler, what is the value of m for which the square number m^2 is a triangular number $n(n + 1)/2$, with $50 < n < 500$. The only answer is $m = 204$, when $n = 288$. This was the answer Mahalanobis was looking for. But he was totally stunned when Ramanujan, stirring his vegetables on a gas fire, answered there is no need of any restriction on the values of n. He rattled out the continued fraction (1.8) and said you can get all the infinite values of m from the rational convergent of this!

When Mahalanobis asked how Ramanujan had arrived at this answer, his reply was even more startling. "Immediately I heard the problem it was clear that the solution should obviously (!) be a continued fraction; I then thought which continued fraction? And the answer came to my mind". Nobody, not even Ramanujan himself, can read the mind of a genius like Ramanujan. After seeing (1.8), some people think he mentally calculated the first two values of m, viz. 1 (with $n = 1$, when the sum of the house numbers on either side is 0) and 6 (with $n = 8$, when the sum of the house numbers on either side is 15). Using these two values 1 and 6 the flash of the intuition of *genius* sparked and the answer was obtained in terms of (1.8). In Appendix A, we show some connection between the solutions of (1.7) and the continued fraction (1.8).

Chapter 2

Integers

2.1 Representation

In most languages numbers are written using Hindu-Arabic numerals. The decimal system is commonly used but the binary system is used in computers. The place value system encountered in both is most convenient and has contributed enormously in the growth of mathematics. In the former, powers of 10 with 10 different digits, viz. $0, 1, 2, \ldots, 9$ are used; whereas in the latter, powers of 2 with only two different digits, viz. 0 and 1 are used. We may recall that ancient civilizations used numbers different from 10 and 2. For example, in Babylon sexagesimal (base 60) system was used. The remnants of these still exist in our time units, like 60 seconds a minute or 60 minutes in an hour. It may be mentioned that there is nothing special about the number 10, except a natural reason that we have 10 fingers which are used in counting. The number 2 for computers is a natural choice as we can define two states of a switch as *On* or *Off*. We discuss these two systems separately.

In the decimal system, the first ten integers are $0, 1, 2, \ldots, 9$. These are alternately called as *even* and *odd,* both digits and integers (whole numbers). Any integer I is expressed by a series of powers of 10 as

$$I = \sum_{i=0}^{n} a_i 10^i \tag{2.1}$$

where a_i is one of the 10 digits with the restriction $a_n \neq 0$. The integer is written using the place value system simply as

$$I = a_n a_{n-1} \ldots a_2 a_1 a_0 \tag{2.2}$$

with $a_n \neq 0$.

Starting from the right end, the values of a_i's give the digits at the places of units, tens, hundreds and so on.

Similarly, in the binary system, an integer I is expressed by a series of powers of 2 as

$$I = \sum_{i=0}^{n} a_i 2^i \tag{2.3}$$

where a_i is one of the two digits 0 or 1 with the restriction $a_n \neq 0$. Thus, a number written in binary system is a string of 1's and 0's. In both (2.1) and (2.3), higher the value of n, higher is the integer.

There is nothing special about the numbers 10 or 2 to be taken as the base for expressing the integers. One can use any number m as the base and then the $(m-1)$ digits to be used are given by $0, 1, 2, \ldots, (m-1)$. A number expressed with some base can easily be expressed with some other base. For example, any integer written in say decimal system can be easily written with any other base. This is explained below by writing a number, for example 893 (in decimal system) in base 5. For this, the given number is successively divided by 5 until the quotient becomes 0. The quotient and remainder are listed as shown below.

Number	Quotient	Remainder
893	178	3
178	35	3
35	7	0
7	1	2
1	0	1

Then the remainders are listed side by side from the last step to the first to write the number in base 5.

Hence, we get $(12033)_5 = 893$. Now we verify that that $(12033)_5$ is really 893 in decimal system. Towards this end, we use the power series of 5 and write the $(12033)_5$ as

$$= 3 \times 5^0 + 3 \times 5^1 + 0 \times 5^2 + 2 \times 5^3 + 1 \times 5^4$$
$$= 3 + 15 + 0 + 250 + 625 = 893$$

Just like the use of sexagesimal system in expressing time, historically we continue to use the roman numerals like I, II, III, IV, V, V, VI, ..., X, ..., etc. in clocks and watches and also in expressing the classes in schools. In this system L, C, D and M are, respectively, used for the numbers 50, 100, 500 and 1000.

In this book, mostly the decimal system will be used, unless otherwise stated. The integers can be geometrically represented by equidistant points on a straight line, as shown in Fig. 2.1. This line is called the (real) number line. From the starting point A representing 0, the line extends indefinitely in both directions.

The positive integers are indicated by points B, C, D, etc. on the right of A and the negative integers are indicated by the points E, F, G, etc. on the left of A.

Fig. 2.1: Integers on the number line

2.2 Test of divisibility

In answering some problems on numbers it is useful to know whether a given number is divisible by a smaller number without carrying out the complete division process. In this section, we list the conditions which decide whether a given number is divisible by numbers from 2, 3, 4, ... up to 11. Of course these results can also be used cleverly to determine divisibility by the product of some of these numbers which are co-prime numbers (i.e. without a common factor).

(i) If the digit in the place of units is even, the number is called *even* and it is divisible by 2.

(ii) If the two-digit number formed by using the last two digits of the given number is divisible by 4, then the number is also divisible by 4 as all hundreds are divisible by 4.

(iii) Extending the logic of (ii), if the three-digit number formed by the last three digits of the given number is divisible by 8, then the number is also divisible by 8 as all thousands are divisible by 8.

(iv) All multiples of 5 end with either 5 or 0. Hence, any number having either 5 or 0 at its units place is divisible by 5.

(v) From (i) and (iv) it is easy to conclude that any number ending with 0 is divisible by 10.

(vi) It may be noted that all integral powers of 10 when divided by 3 (or 9) leave a remainder 1. So any integer expressed in decimal system by (2.1) when divided by 3 (or 9), the remainder will be $\sum_{i=0}^{n} a_i$. So if this sum of the (coefficient) digits of a given number is divisible by 3 (or 9), then the number will be divisible by 3 (or 9).

(vii) From (i) and (vi) it is easy to conclude that if an even number is divisible by 3, then it is divisible by 6.

(viii) We note that 10^n if divided by 11, then the remainder is 10 if n is odd and the remainder is 1 if n is even. So an integer I given by (2.1) when divided by 11 leaves a remainder

$$R = (a_0 + a_2 + a_4 + \dots) + 10(a_1 + a_3 + a_5 + \dots)$$
$$= (a_0 + a_2 + a_4 + \dots) + (11 - 1)(a_1 + a_3 + a_5 + \dots)$$
$$= \underbrace{\sum a_k}_{k=\text{even}} - \underbrace{\sum a_l}_{l=\text{odd}} + 11p$$

with p an integer. Hence, the integer I is divisible by 11 if the difference of the sums of alternate digits is divisible by 11. It should be remembered that 0 is divisible by all numbers.

It must have been noticed that so far we have covered checking the divisibility by all the numbers from 2 to 11, except 7. No convenient divisibility check for 7 is included in usual textbooks. Historians of mathematics have found a divisibility test for 7 which is more than 1500 years old. In 2018 a very similar method was discovered by a student of class 7, who noticed a curious property of all multiples of 7. His name is Chika Ofili. He is from Nigeria and studies in a school in England. Based on his observation, we can devise a method for checking the divisibility of a given number by 7. But this method is not a one step process like all other numbers outlined in (i) to (viii). Here the size of the given number is reduced at every step and finally reaches a small number which is easy to check for its divisibility by 7. It is not difficult to prove that Chika's observation is true for all multiples of 7. Now we first mention Chika's observation and then prove its general validity. This discussion is included to show that by keen observation and enthusiasm even a school student can discover new mathematical truths, for which a rigorous mathematical proof can be obtained later.

Let us consider the multiples of 7, viz. $14, 21, 28, 56, 63, \dots$. Chika first multiplied the units digit by 5 and added this product to the number obtained by deleting the units digit. He observed that the latter sum is also divisible by 7. This condition is both necessary and sufficient as is shown shortly. First, as examples, let us consider 14, 49, 77, 126, 196, and 40124, as some multiples of 7. Applying Chika's procedure to these numbers, we get respectively $21(= 1 + 5 \times 4), 49(= 4 + 5 \times 9), 42(= 7 + 5 \times 7), 42(= 12 + 5 \times 6), 49(= 19 + 5 \times 6), 4032(= 4012 + 5 \times 4)$. It is easy to see that the first 5 numbers are multiples of 7. To check for the last number, we continue to apply Chika's procedure and get $413(= 403 + 5 \times 2)$, which in turn, gives $56(= 41 + 5 \times 3)$ which is easily seen to be a multiple of 7. It should be noted that in each step no division process is employed. Chika tried with many large numbers and found no exception and also for the converse that if the original number is not a multiple of 7, then the number obtained following his procedure is also not a multiple of 7. In what follows we prove the general validity of these statements without referring to any particular number.

First we write any general integer using (2.1) as

$$I = a_n a_{n-1} a_{n-2} \ldots a_2 a_1 a_0 = 10T + a_0 \tag{2.4}$$

where $T = a_n a_{n-1} a_{n-2} \ldots a_3 a_2 a_1$.

Now we prove that: (a) If the number $T + 5a_0$ is divisible by 7, then so is the number I; (b) If the number I is divisible by 7, then so is the number $T + 5a_0$; (c) If the number I is not divisible by 7, then the number $(T + 5a_0)$ is also not divisible by 7.

(a) Let us write $T + 5a_0 = 7m$, where m is an integer; so we can write

$$I = 10(7m - 5a_0) + a_0 = 7(10m - 7a_0) \tag{2.5}$$

Therefore, I is divisible by 7.

(b) Let us write $I = 10T + a_0 = 7p$, where p is an integer. So we can write

$$5a_0 + T = 5(7p - 10T) + T = 7(5p - 7T) \tag{2.6}$$

Therefore, the number $(T + 5a_0)$ is divisible by 7.

(c) Let us write

$$I = 7n + r \tag{2.7}$$

where $r = 1, 2, 3, 4, 5$ or 6, and n is an integer.

Using (2.7) in (2.4), we write

$$a_0 = 7n + r - 10T \tag{2.8}$$

Using (2.8), we get

$$5a_0 + T = 5(7n + r - 10T) + T = 7(5n - 7T) + 5r \tag{2.9}$$

It is easily seen that for all possible values of r, the number $5r$ and hence the number $5a_0 + T$ is not divisible by 7.

We may conclude this section by mentioning that in higher arithmetic, similar tests for divisibility by 13 and 17 are available. Enthusiastic students can explore further.

2.3 Curious patterns

In this section, we discuss some curious patterns formed by arithmetical operations, like multiplication, addition, squaring, etc. on integers. Only the operations are indicated and the task of observing or extending the patterns, whenever possible, is left for the readers.

2.3.1 Multiplication

(a)
$$1 \times 1 = 1$$
$$11 \times 11 = 121$$
$$111 \times 111 = 12321$$
$$1111 \times 1111 = 1234321$$
$$11111 \times 11111 = 123454321$$
$$\vdots$$
$$111111111 \times 111111111 = 12345678987654321$$

(b)
$$11 \times 91 = 1001$$
$$11 \times 9091 = 100001$$
$$11 \times 909091 = 10000001$$
$$\vdots$$
$$11 \times 9090909090909091 = 100000000000000001$$

2.3.2 Multiplication, addition and division

(a)
$$1 \times 8 + 1 = 9$$
$$12 \times 8 + 2 = 98$$
$$123 \times 8 + 3 = 987$$
$$1234 \times 8 + 4 = 9876$$
$$\vdots$$
$$123456789 \times 8 + 9 = 987654321$$

(b)
$$0 \times 9 + 1 = 1$$
$$1 \times 9 + 2 = 11$$
$$12 \times 9 + 3 = 111$$
$$123 \times 9 + 4 = 1111$$
$$\vdots$$
$$12345678 \times 9 + 9 = 111111111$$

(c)
$$0 \times 9 + 8 = 8$$
$$9 \times 9 + 7 = 88$$
$$98 \times 9 + 6 = 888$$
$$987 \times 9 + 5 = 8888$$
$$\vdots$$
$$98765432 \times 9 + 0 = 888888888$$

(d)
$$(8 \times 8) + 13 = 77$$
$$(8 \times 88) + 13 = 717$$
$$(8 \times 888) + 13 = 7117$$
$$\vdots$$
$$(8 \times 88888888) + 13 = 711111117$$

In such pattern formations, sometimes the style of presentation, rather than the underlying mathematics, may be the root cause of the curious pattern. For example, in the example to follow, the pattern does not appear to be mysterious if we observe that $111 = 37 \times 3$ and multiplying the dividend and the divisor by the same factor leaves the quotient unaltered.

(e) $\dfrac{111}{1+1+1} = \mathbf{37}$ $\dfrac{222}{2+2+2} = \mathbf{37}$ $\dfrac{333}{3+3+3} = \mathbf{37}$

$\dfrac{444}{4+4+4} = \mathbf{37}$ $\dfrac{555}{5+5+5} = \mathbf{37}$ $\dfrac{666}{6+6+6} = \mathbf{37}$

$\dfrac{777}{7+7+7} = \mathbf{37}$ $\dfrac{888}{8+8+8} = \mathbf{37}$ $\dfrac{999}{9+9+9} = \mathbf{37}$

2.3.3 With consecutive integers

(a)
$$1 + 2 = \mathbf{3}$$
$$4 + 5 + 6 = \mathbf{7 + 8}$$
$$9 + 10 + 11 + 12 = \mathbf{13 + 14 + 15}$$
$$16 + 17 + 18 + 19 + 20 = \mathbf{21 + 22 + 23 + 24}$$
$$\vdots$$

This pattern with consecutive integers continues forever.

(b)
$$3^2 + 4^2 = \mathbf{5^2} \text{ (Also see Section 2.8)}$$
$$10^2 + 11^2 + 12^2 = \mathbf{13^2 + 14^2}$$
$$21^2 + 22^2 + 23^2 + 24^2 = \mathbf{25^2 + 26^2 + 27^2}$$
$$36^2 + 37^2 + 38^2 + 39^2 + 40^2 = \mathbf{41^2 + 42^2 + 43^2 + 44^2}$$
$$\vdots$$

You may observe and prove that the first integer of the kth row is given by $(2k^2 + k)^2$.

(c) Mathematicians have shown that the sum of the squares of 24 consecutive integers will be a perfect square, only if the starting number is 1, 9, 20, 25, etc. as shown below.
$$1^2 + 2^2 + 3^2 + \cdots + 23^2 + 24^2 = \mathbf{70^2}$$
$$9^2 + 10^2 + 11^2 + \cdots + 31^2 + 32^2 = \mathbf{106^2}$$
$$20^2 + 21^2 + 22^2 + \cdots + 42^2 + 43^2 = \mathbf{158^2}$$

and so on. It has been also seen that none of the sums of the squares of up to the first 23 squares, i.e. $(1^2 + 2^2), (1^2 + 2^2 + 3^2), \ldots, (1^2 + 2^2 + 3^2 + \cdots + 23^2)$ is a perfect square.

(d) It can be easily seen that $3^3 + 4^3 + 5^3 = \mathbf{6^3}$.

But unlike in section (b) above, with squares rather than cubes, this is more of an exception; the pattern cannot be extended any further.

2.3.4 Pascal's triangle

Blaise Pascal (1623–62) was a French genius. Some integers can be written in a triangular pattern called *Pascal's Triangle*. He popularized this triangle by applying it to solve problems in probability. This triangle, discovered by many other mathematicians, is known by different names in different countries. For example,

Pingala in India (sometime between 500 to 300 BC) mentioned this triangle in his famous book Chandahsastra. In Iran this triangle is credited to Omar Khyyam (1048–1131), in China to Liu Hui (in the third century), and in Italy to Tartaglia (1499–1557).

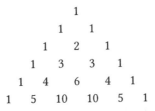

Fig. 2.2: Pascal's triangle

The first six rows of the Pascal's triangle are shown in Fig. 2.2. The Pascal's triangle can be extended for ever. We can make this triangle following several easy and simple rules. Only one of these is mentioned here. We start with one 1 at the top row. Then we write in each successive row as many numbers as the row number, so as to make an overall equilateral triangle at every step; i.e. two numbers in the second row and three numbers in the third row and so on. The numbers in each row are written as follows. Go to the previous row above the number to be written, find the numbers to the left and right of this imagined location and write down the sum of these two numbers. If no number is found either on the left or on the right, then consider that number as zero. The reader is encouraged to make the Pascal triangle by following this rule and verify that Fig. 2.2 is correct.

It may be noted that the sum of the entries of the kth row is 2^{k-1}. Further, starting from the second row onwards, if we put alternately plus and minus signs before each entry in a row and take the sum, then we get zero. This statement may seem to be obvious for the even numbered rows due to symmetrical entries of equal numbers about an imaginary central line. However, the statement remains valid also for the odd numbered rows (except the first one, where there is a single number) without such symmetrical entries of equal numbers. We will see an application of this observation when we discuss Bernoulli's numbers in Chapter 3.

The Pascal's triangle finds a number of applications in various branches of mathematics. Here we mention some very simple ones. Let us count the row number n starting from the second one (i.e. $n = 0$ for the top row) and the element number k in each row again with $k = 0$ for the first number. Then the kth element in the nth row gives the value of n_{c_k}, also written as $\binom{n}{k}$, which signifies the number of different teams consisting of k members that can be made from a total number of n persons, with $n > k$. Mathematically this is called the *possible number of combinations* of n different things taken k at a time. Counting the number n in the manner mentioned above, we may observe that the entries in the nth row give the coefficients of the binomial expansion $(1 + x)^n$.

Now we shift the numbers of the Pascal's triangle (Fig. 2.2) horizontally so that the first entry of each row falls on the same vertical line, as shown in Fig. 2.3. It is observed that if, in this figure, starting from the first number in each row, we add up all the numbers encountered by moving diagonally upward, we get the sequence of numbers $(1, 1, 2, 3, 5, 8, 13, \ldots)$. This is a very famous sequence, known as *Fibonacci (Hemachandra) sequence,* which we will discuss later in Section 2.9. Moreover, by writing the second number in each row (starting from the second row) in Fig. 2.2 or 2.3, we get the sequence of natural numbers $1, 2, 3, 4, \ldots$ and by writing the third number in each row, starting from the third row, we get the sequence of triangular numbers $1, 3, 6, 10, \ldots$.

$$
\begin{array}{cccccc}
1 \\
1 & 1 \\
1 & 2 & 1 \\
1 & 3 & 3 & 1 \\
1 & 4 & 6 & 4 & 1 \\
1 & 5 & 10 & 10 & 5 & 1
\end{array}
$$

Fig. 2.3: Left-aligned shifted Pascal's triangle

2.4 Iterations

The process of successively applying an identical set of arithmetical operations on a given integer is called *Iteration.* Starting from an initial value the outcome at every stage, called the *iterate,* might show interesting behavior. The behavior depends both on the set of defined operations and the initial value. Here we discuss a few interesting iterations. Some obvious and totally expected behavior such as continually increasing by squaring the number at every stage, starting with an integer greater than 1, is not interesting. The ultimate behaviour of the iterates is called an *attractor* and the set of initial values resulting in the same attractor is called the *basin of attraction* of that attractor. The attractor may indicate that the iterates finally arrive at a fixed value; in that case the attractor is referred to as a *fixed point attractor.* Sometimes the iterates are seen to be periodically wandering across a set of values. In such a situation, the attractor is called an *n-cycle periodic attractor,* where n is the number of elements in the set defining the periodic attractor. The examples given below will clarify these concepts and definitions.

2.4.1 Number of even, odd and total digits

Let us start from an arbitrary integer. The iteration process is defined as follows. Count the number of even, odd and the total digits in this integer and write these numbers side by side to produce the iterate. Then continue the same process. Starting from any integer a fixed point attractor 123 results. As an example, let us start with 57, 681, 923, 403, 481, 610, 583, 257. It can be seen that this integer has 11 even (remember 0 is an even digit), 12 odd and 23 total digits. So the first iterate is 111223. Continuing we get 246 → 303 → 123 → 123 → . . . . Thus, we have arrived at the fixed point attractor 123. You may start with a different starting value and convince yourself that all integers constitute the basin of attraction for this fixed point attractor of this iteration.

2.4.2 Sum of powers of the digits

First consider that starting from a positive integer the iterate produces the sum of the squares of the digits, and then the process continues. It has been proved that for this iteration, both types of attractors result, each one having its own basin of attraction. The periodic attractor is always an 8-cycle periodic attractor, given by $89 \rightarrow (8^2 + 9^2) \rightarrow 145 \rightarrow 42 \rightarrow 20 \rightarrow 4 \rightarrow 16 \rightarrow 37 \rightarrow 58 \rightarrow 89 \rightarrow \dots$.

For most of the initial values, within a few steps the iterate becomes one of the above eight values and then follows this periodic attractor. At which of the above values the iterate hits for the first time depends on the initial value. You may verify by starting with two different initial values, such as 25 and 26. For some typical initial values such as 68, we get $68 \rightarrow 100 \rightarrow 1 \rightarrow 1 \rightarrow \dots$, i.e. a fixed point attractor 1. It has been observed that the basin of attraction of the periodic attractor is much larger than that of the fixed point attractor.

At this stage we leave a small exercise for the readers. It is known that 153, 370, 371 and 407 are the only four numbers (not counting the trivial value 1) which are sum of the cubes of their digits. Out of these only 153 is divisible by 3. Now verify that starting from any multiple of 3 (say 63), if the iterate produces the sum of the cubes of the digits and then the process continues, one reaches a fixed point attractor 153. The number of steps needed to hit the fixed point attractor obviously depends on the starting value.

2.4.3 Magic show with 1089

You can always show number magic to your friend. You ask him to write secretly a three-digit number with different digits at the places of units and hundreds. Then ask him to reverse the digits of the first to get a second number. Ask him to obtain the positive difference of these two numbers (i.e. subtract the smaller one from the larger of these two numbers). Now ask him to reverse the digits of the number so obtained and then add these last two numbers. You announce this sum as 1089 totally independent of the number your friend has written. As an example, let the number written by your friend be 493. Then the second number is 493 − 394 = **099**. Reversing the digits you get 990 and adding this to 099, one gets 990 + 099 = **1089**. It is easy to prove why is it always 1089 independent of the number written by your friend. You are invited to furnish the proof. If you are unable to do it, then you can see the proof at the end of Section 2.4.

2.4.4 Kaprekar numbers

Two numbers, viz. 495 and 6174, are called *Kaprekar numbers* in honour of Datta-treya Ramachandra Kaprekar (1905–86), an Indian recreational number theorist, who discussed these two attractors for two different iterations. In the first one, a three-digit number with at least two different digits is the starting number. Then we write the largest and smallest numbers using these digits and take the positive difference to generate the first iterate. If we continue this iteration, finally, we get the fixed point attractor 495. Exactly the same way, starting with a four-digit number (with the restriction that not all digits are same) the iteration now produces a fixed point attractor 6174. The proof of these attractors is a little more involved than that of Section 2.4.3, so will be omitted. Below we give one example of each of the iteration starting, respectively with 313 and 1767.

With initial value 313, we get the largest and smallest numbers using these digits (i.e. with two 3's and one 1) as 331 and 133. Thus, we can write the following iterates:

$$313 \rightarrow 133 = 180 \rightarrow 810$$
$$= 792 \rightarrow 972 - 279$$
$$= 693 \rightarrow 963 - 369$$
$$= 594 \rightarrow 954 - 459$$
$$= 495 \rightarrow 954 - 459$$
$$= 495 \ldots$$

Similarly, starting with 1767, the largest and smallest numbers using these digits are obtained as 7761 and 1677, respectively. Thus, we can write the following iterates:

$$1767 \rightarrow 7761 - 1677 = 6084 \rightarrow 8640 - 0468$$
$$= 8172 \rightarrow 8721 - 1278$$
$$= 7443 \rightarrow 7443 - 3447$$
$$= 3996 \rightarrow 9663 - 3699$$
$$= 6264 \rightarrow 6642 - 2466$$
$$= 4176 \rightarrow 7641 - 1467$$
$$= 6174 \rightarrow 7641 - 1467 \rightarrow 6174 \ldots$$

2.4.5 Collatz conjecture and hailstone numbers

Let an iteration be defined through the following equation:

$$x_{k+1} = \begin{cases} 3x_k + 1, & \text{if } x_k \text{ is odd} \\ (x_k/2), & \text{if } x_k \text{ is even.} \end{cases} \tag{2.10}$$

Lothar Collatz (1910–90) conjectured in 1937 that this iteration exhibits a 3-cycle periodic attractor $4, 2, 1, 4, 2, 1, \ldots$ independent of the initial value. This conjecture has still not been proved but no exception has been found for any initial value in the range 1 to 5.764×10^{18}. As an example, let us follow this iteration with a starting value 17 as shown below:

$$19 \rightarrow 58 \rightarrow 29 \rightarrow 88 \rightarrow 44 \rightarrow 22 \rightarrow 11 \rightarrow 34 \rightarrow 17 \rightarrow 52 \rightarrow 26 \rightarrow 13 \rightarrow 40$$
$$\rightarrow 20 \rightarrow 10 \rightarrow 5 \rightarrow 16 \rightarrow 8 \rightarrow 4 \rightarrow 2 \rightarrow 1 \rightarrow 4 \rightarrow 2 \rightarrow 1 \ldots$$

The number of iterates needed to hit the periodic attractor depends on the initial value. For example, if you start with 27 it will take 111 iterations to reach 1 for the first time. We know that in a storm cloud, the hailstones drift haphazardly upward and downward before ultimately hitting the ground. It can be easily seen that the iterated value shows a similar characteristic, i.e. haphazardly increases and decreases depending on whether the current value is odd or even; but ultimately reaches the value 1. Consequently, the iterated numbers are called the *hailstone numbers*. A lot of computer study has been made with this iteration and a hailstone number as high as 9,232 has been reached. With the starting value 63,728,127 it takes 949 iterations to reach 1.

[*Proof of Section* 2.4.3: Let the number written by your friend be *abc*. Then reversing the digits the friend writes *cba*. Now $abc = 100a + 10b + c$, similarly $cba = 100c + 10b + a$. The positive difference of these two numbers is $99|a - c| = 100(|a - c| - 1) + 10 \times 9 + (10 - |a - c|)$. Reversing the digits of this number, we get the number $100(10 - |a - c|) + 10 \times 9 + (|a - c| - 1)$. Adding these last two numbers we obtain $1000 - 100 + 2 \times 10 \times 9 + 9 = 900 + 180 + 9 = 1089$, independent of *a, b* and *c*.]

2.5 Prime numbers

All of you are aware of the simplest classification of integers into odd and even categories. Another very important basic classification of numbers is in two groups called *prime* and *composite*. Prime numbers are those which do not have any divisor other than 1 and the number itself. The first few prime numbers are $2, 3, 5, 7, 11, 13, 17, \ldots$. By convention, the number 1 is not listed as a prime. The reason for this will be explained soon. It must be noted that 2 is the only even prime. The prime numbers are as important to arithmetic as the atoms are to the physics of materials.

The fundamental theorem of arithmetic states that all composite numbers can be written as a product of its prime factors in a unique way, but for their order. For example, we can write $18 = 2 \times 2 \times 3$ (or, $2 \times 3 \times 2$, etc.). No new prime can appear in the factorization of the number 18. The prime factorization is unique. The mathematical proof of this apparently obvious, but very important and useful, theorem is not easy and will not be discussed. To protect this theorem, 1 is not included as a prime number. With the inclusion of 1 as a prime, the prime factorization of a composite number will not remain unique. With 1 also as a prime we could have put 1 in the prime factorization any arbitrary number of times. We could have written $18 = 1 \times 2 \times 1 \times 2 \times 3$ also. Because 1 is not defined as a prime, we never include 1 in the prime factorization, and thus maintain the uniqueness of prime factorization.

Mathematicians are working on prime numbers for more than 2000 years. A lot of theorems and results have been obtained. Still some interesting conjectures, neither disproved nor rejected by counterexample, are waiting to be proved. Renewed interest in prime numbers in the modern age is due to the fact that the transfer of secret coded messages through internet depends on some results based on prime numbers. Both serious and recreational mathematicians are involved with prime numbers. Prime numbers are infinite; some of the prime numbers with special characteristics or forms are named after great mathematicians. In this section, we discuss briefly some interesting and useful results on prime numbers.

2.5.1 Test of primality

Carl Friedrich Gauss (1777–1855), considered as one of the greatest mathematicians in history, said that two important questions in number theory are to answer how to determine whether a given number, n, is a prime or not and if not, then what are its prime factors. The simplest solution of the first question is to check whether n is divisible by all the prime numbers up to $(n)^{1/2}$ or not. If not, then n is a prime. However, this method is not suitable for large numbers as the computation time increases at a very fast rate as the number increases. This increase in time requirement renders the simple method not practicable. Since then efficient algorithms for testing the primality of a given number have been created and are being continuously improved.

Greek mathematician Eratosthenes (276–194 BC) devised a method to detect the primes, known as the *Eratosthenes sieve*. This apparently simple method has been used to obtain some nice mathematical results. So we discuss this *sieve* before talking about the application.

First, all the numbers starting from 2 are written one after another as shown in Fig. 2.4.

```
  2   3   4   5   6   7   8   9  10  11  12  13  14  15
 16  17  18  19  20  21  22  23  24  25  26  27  28  29
 30  31  32  33  34  35  36  37  38  39  40  41  42  43
 44  45  46  47  48  49  50  51  52  53  54  55  56  57
 58  59  60  61  62  63  64  65  66  67  68  69  70  71
 72  73  74  75  76  77  78  79  80  81  82  83  84  85
 86  87  88  89  90  91  92  93  94  95  96  97  98  99
100 101 102 103 104 105 106 107 108 109 110 111 112 113
```

Fig. 2.4: Starting of Eratosthenes sieve

Then all the multiples of 2, like 4, 6, 8, 10, . . . , etc. are erased and we get Fig. 2.5. Then from this figure we remove all the existing multiples of 3, (greater than 3) and obtain Fig. 2.6. In the next step we remove all the existing multiples of 5, (greater than 5) and get Fig. 2.7. It is easy to see that in Fig. 2.7 all the prime numbers up to 47 have been sieved out. This process of removing the existing multiples of sieved out primes is continued to sieve out more and more prime numbers. In the next step all prime numbers up to 113 will be sieved out, removing multiples of 7, like 77 and 91.

```
 2   3   .   5   .   7   .   9   .  11   .  13   .  15
 .  17   .  19   .  21   .  23   .  25   .  27   .  29
 .  31   .  33   .  35   .  37   .  39   .  41   .  43
 .  45   .  47   .  49   .  51   .  53   .  55   .  57
 .  59   .  61   .  63   .  65   .  67   .  69   .  71
 .  73   .  75   .  77   .  79   .  81   .  83   .  85
 .  87   .  89   .  91   .  93   .  95   .  97   .  99
 . 101   . 103   . 105   . 107   . 109   . 111   . 113
```

Fig. 2.5: Eratosthenes sieve with multiples of 2 removed

```
 2   3   .   5   .   7   .   .   .  11   .  13   .   .
 .  17   .  19   .   .   .  23   .  25   .   .   .  29
 .  31   .   .   .  35   .  37   .   .   .  41   .  43
 .   .   .  47   .  49   .   .   .  53   .  55   .   .
 .  59   .  61   .   .   .  65   .  67   .   .   .  71
 .  73   .   .   .  77   .  79   .   .   .  83   .  85
 .   .   .  89   .  91   .   .   .  95   .  97   .   .
 . 101   . 103   .   .   . 107   . 109   .   .   . 113
```

Fig. 2.6: Eratosthenes sieve with multiples of 2 and 3 removed

Euclid (325–265 BC) gave a simple proof that the prime numbers never end, there is nothing like the largest prime number. He proved it in a simple manner by using the principle of contradiction. If there is a largest prime, say the *n*th prime, denoted by p_n, then consider the number P, which is one more than the product of all the primes, i.e.

$$P = p_1 p_2 p_3 \ldots p_n + 1 \tag{2.11}$$

It is obvious that P is not divisible by any prime number up to p_n; this gives rise to two possibilities, (i) P is a prime or (ii) P, if composite, has a prime factor greater than p_n. As P is greater than p_n, both possibilities imply p_n cannot be the largest prime.

Hence, the non-existence of the largest prime is proved.

So we now know that both the set of the natural numbers and prime numbers have infinite elements. Using the technique of Eratosthenes sieve, Euler obtained the following equation which provides a very useful relationship between these two universes of infinities:

$$\sum_n \frac{1}{n^s} = \prod_p \frac{1}{\left(1 - \frac{1}{p^s}\right)} \tag{2.12}$$

where *n* refers to the natural numbers and *p* refers to the prime numbers, *s* is a given number greater than 1. On the left of the above equation is a sum of infinite series as *n* takes the values $1, 2, 3, \ldots$ and on the right is a product of infinite number of terms as *p* takes the values of all the prime numbers. This, relation is of immense importance in various branches of mathematics and is known as *The Golden Key.*

2.5.2 Types of primes

Some primes with special properties are given specific names, mostly after the great mathematicians whose works defined those. Here we just mention only a few of them.

(a) *Euclidean Primes*: If the number P defined in eq. (2.11) is a prime then such a prime is called *Euclidean Prime*. The first few Euclidean primes are 3, 7, 31, 211, 2311, ...

(b) *Mersenne Primes*: These primes are named after Marin Mersenne (1588–1648). If with a prime p, the number $2^p - 1$ is also another prime, then the latter is called *Mersenne prime*. You can easily prove that with a non-prime n, $2^n - 1$ cannot be a prime. The first few Mersenne primes are $3, 7, 31, 127, 8191, \ldots$, (respectively with $p = 2, 3, 5, 7, 13, \ldots$). Till 2018 only 50 Mersenne primes have been found. The last one, for $n = 72, 232, 917$ having 23, 249, 425 digits was the largest known prime till that time.

(c) *Fermat Primes*: Mathematicians always tried to generate prime numbers by some formula, if not all the primes, at least only primes. Pierre de Fermat (1601–65) claimed to have obtained such a formula as

$$F_p = 2^{2^p} + 1 \tag{2.13}$$

where p represents non-negative integers. The first few values of F_p are shown in Table 2.1.

Table 2.1

p	F_p
0	3
1	5
2	17
3	257
4	65, 537
5	4, 294, 967, 297

Initially F_p's were known as *Fermat primes*. But later on (long after the death of Fermat), Euler showed that $F_5 = 641 \times 6{,}700{,}417$ and hence not a prime. In fact, now F_p's are called *Fermat numbers*. The first five numbers in Table 2.1 are Fermat primes.

Surprising connections are the most beautiful things in mathematics. We talk of one such connection here. Greek geometers knew how to geometrically draw an equilateral triangle and a regular pentagon. But they failed to draw a regular polygon with the number of sides given by the next prime number 7. Gauss showed that geometrical construction of regular polygons with prime number of sides is possible only for Fermat primes. He showed how to draw a regular polygon with 17 sides. Later on someone drew a regular polygon with 257 sides.

(d) *Sophie Germain* (1776–1831) *Primes*: A prime p, for which the number $(2p+1)$ is also a prime, is called *Sophie Germain prime*. The first few Sophie Germain primes are $(2, 3, 5, 11, 23, \ldots)$.

It is not known whether any of the three categories (a), (b) and (d) of primes continue forever or not.

(e) *Ramanujan Primes*: Ramanujan prime R_n is the smallest integer for which

$$\pi(x) - \pi\left(\frac{x}{2}\right) \geq n \text{ for all } x \geq R_n \tag{2.14}$$

where $\pi(x)$ is called the *prime counting function* giving the number of primes up to and including x. The first few Ramanujan primes are 2, 11, 17, 29, 41, ... (respectively with $n = 1, 2, 3, 4, 5, \ldots$). Equation (2.12) implies $\pi(R_n) - \pi(R_n/2) = n$.

(f) *Pillai Primes*: These are named after the great Indian mathematician Subbayya Sivasankaranarayana Pillai (1901–50). A prime number p for which there is a positive integer n such that $n!$ is 1 less than a multiple of p, but p is not 1 more than a multiple of n is called a *Pillai prime*. Before writing this definition mathematically, this is the right place to introduce a one-way function which will be used in this book. The one way function called *mod function*; mod($\cdot$) is defined as $b = a \bmod (n)$ implying that b divided by n leaves a remainder a. It should be noted that the function is one way in the sense for given values of b and n, we can uniquely determine a, but not the other way round, i.e. with given values a and n, we cannot determine b uniquely. One can only say that $b = kn + a$, where k is any integer. Now we can mathematically write a Pillai prime P_p such that the following conditions are satisfied:

$$n! = -1 \bmod (P_n), \text{ but } P_p \not\equiv 1 \bmod (n) \qquad (2.15)$$

The first few Pillai primes are 23, 29, 59, 61, ... You can verify that $n = 14$ for the first Pillai prime as $n! = 87, 178, 291, 200$. Both Ramanujan and Pillai primes continue forever.

(g) *Twin Primes*: Except 2 all primes are odd. If two consecutive odd numbers are primes, then that pair is called *twin primes*. Examples of twin primes are $(3; 5)$, $(5; 7)$, $(11; 13)$, $(17; 19)$, ... $(824, 633, 702, 441; 824, 633, 702, 443)$, ... No one knows whether twin primes continue forever or not. However, it has been proved, in 1919, by Viggo Brun (1885–1978) that the sum of the series of the reciprocals of twin primes, i.e.

$$S = \left(\frac{1}{3} + \frac{1}{5} \right) + \left(\frac{1}{5} + \frac{1}{7} \right) + \left(\frac{1}{11} + \frac{1}{13} \right) + \cdots \qquad (2.16)$$

converges. Since an infinite series may also converge, the question whether twin primes continue forever or not is not settled. Had it diverged we could have said that twin primes continue forever. In 1994, while obtaining the approximate value of this Brun's constant ($= 1.9021605824\ldots$) a *bug* in the FPU (floating point unit) of the Intel Pentium processor was detected while calculating the reciprocals of the twelve-digit twin primes mentioned above. Next year all the processors all over the world had to be replaced at a cost of \$ 475 million.

We just discussed Fermat's attempt and failure to generate only prime numbers by a formula. Great mathematicians like Euler and Adrien-Marie Legendre (1752–1833) created the following prime generating polynomials

$$n^2 - n + 41 \qquad (2.17)$$
$$\text{and } n^2 - 79n + 1601 \qquad (2.18)$$

with n representing non-negative integers. It is found that (2.17) generates 40 prime numbers when $n = 0, 1, 2, \ldots, 40$ and (2.18) generates 80 prime numbers with $n = 0, 1, 2, \ldots, 79$. Both formula fail when they generate 41×41 with $n = 41$ and $n = 80$, respectively.

Mathematicians have tried in vain to find some pattern in the appearance of prime numbers in the sequence of natural numbers. No one could give the answer to the question *Given an arbitrary number n, when will the next prime number appear?* Gauss changed the question to *Given an arbitrary number n, how many prime numbers have appeared up to n?* Gauss proposed such a prime counting function [see eq. (2.14)] which we will briefly discuss in Chapter 3.

We end this section by stating two famous unproven conjectures on prime numbers, one made in 1742 and the other in 1985. Using computers no counter example to these two conjectures has been found so far. In 1742, Christian Goldbach (1690–1764) requested Euler in a letter to prove that "any even number greater than 2 can be expressed as a sum of two primes". This is known as *Goldbach's conjecture*. No counter example has been found up to a number as high as 10^{18}. In 1985, Dorin Andrica (1956–) conjectured that the difference between the square roots of two consecutive prime numbers is less than unity. This is called *Andrica's conjecture*. Mathematically, it is written as

$$\sqrt{p_{n+1}} - \sqrt{p_n} < 1 \tag{2.19}$$

where p_n indicates the nth prime. It has been verified to be true up to n as high as 10^{16}. The highest difference between the square roots of two successive primes has been found to be

$$\sqrt{11} - \sqrt{7} \approx 0.67087 \tag{2.20}$$

Curious facts about prime numbers are galore. One is that you can easily remember a prime number consisting of 3,793 digits! This is written in short as $(1676)_{948}1$. The subscript implies that the number in the parenthesis has to be repeated side by side 948 times before the last digit 1.

2.6   Composite numbers and their divisors

Using the fundamental theorem of arithmetic any composite number C can be written in terms of prime factors as

$$C = p_1^{n_1} \times p_2^{n_2} \times p_3^{n_3} \times \cdots \times p_k^{n_k} \tag{2.21}$$

where p_i is the ith prime number, i.e. $p_1 = 2$, $p_2 = 3$, $p_3 = 5$, ... and n_i's are non-negative integers with $n_k \neq 0$. If a particular prime p_m is absent in the factorization, the corresponding exponent n_m is zero.

A composite number has a number of divisors. The list of divisors of a composite number, C, includes 1, C and all the numbers that divide C without leaving any remainder. Using (2.21) it is easy to determine the total number of such divisors of C. We can pick up one number each from the following sets and their product gives a divisor:

$$(2^0, 2^1, 2^2, \ldots, 2^{n_1}), \ (3^0, 3^1, 3^2, \ldots, 3^{n_2}), \ \ldots, \ (p_k^0, p_k^1, p_k^3, \ldots, p_k^{n_k})$$

If the first element is picked up from every set, then the product gives the divisor 1: similarly, if the last element is selected from each set then the product gives the divisor C.

Hence, the number of divisors

$$D = (1 + n_1)(1 + n_2)\ldots(1 + n_k) \tag{2.22}$$

It may be noted that the special number 1 has only one divisor, viz. 1, any prime number p has two divisors, viz. 1 and p; any composite number of the form p^2, i.e. square of a prime number has 3 divisors, viz. 1, p and p^2.

Special composite numbers

In this section, we mention some composite numbers, each having special characteristics studied by mathematicians.

(a) *Highly composite numbers*: Ramanujan defined and studied these numbers in a very long paper. This work was ultimately submitted to Cambridge University for awarding him the BA degree (by research). He defined a highly composite number as one that has more number of divisors than all other composite numbers less than that. The first few and some large highly composite numbers, their prime factorization and the numbers of divisors are listed below in Table 2.2.

Table 2.2

Highly composite number	*Prime factorization*	*Number of divisors*
6	$2^1 \times 3^1$	4
12	$2^2 \times 3^1$	6
24	$2^3 \times 3^1$	8
36	$2^2 \times 3^2$	9
...	...	...
332, 640	$25 \times 33 \times 5 \times 7 \times 11$	192
43, 243, 200	$26 \times 33 \times 52 \times 7 \times 11 \times 13$	672
2, 248, 776, 129, 600	$26 \times 33 \times 52 \times 72 \times 11 \times 13 \times 17 \times 19 \times 23$	8,064

According to Ramanujan, for all highly composite numbers the exponents of the prime factors obey the following relationship:

$$n_1 \geq n_2 \geq n_3 \geq \cdots \geq n_k \tag{2.23}$$

Further, he also claimed that for all highly composite numbers, except 36, the last exponent is 1. Even without a computer, Ramanujan obtained all the highly composite numbers up to 6, 746, 328, 386, 800. He missed only one such number, viz. 29, 331, 862, 500.

(b) *Perfect numbers*: The *proper divisors* of a composite number are defined as all the divisors including 1 but excluding the number itself. Perfect numbers are those which are the sum of all its *proper* divisors. The first few are 6, 28, 496,

8128, ... You can easily verify that proper divisors of 6 are 1, 2 and 3 and also 6 = 1 + 2 + 3; similarly, proper divisors of 28 are 1, 2, 4, 7, 14 and these add up to 28. The reader can verify this fact for 496 and 8128. Another unexpected connection between two types of numbers is that between the perfect numbers and Mersenne primes. Euler proved Euclid's observation that all even perfect numbers are of the form $2^{n-1} \times (2^n - 1)$, where the number within the parenthesis is a prime number. We may recall that such a prime number is now called a *Mersenne prime* defined in Section 2.5.2. So we know that till 2018, only 50 perfect numbers have been found. No one knows whether a perfect number can be odd or not. Though none has been found but its non-existence has not been proved. It may also be pointed out, that except for 6, all other perfect numbers are sums of cubes of consecutive odd numbers, like $28 = 1^3 + 3^3$; $496 = 1^3 + 3^3 + 5^3 + 7^3$; $8,128 = 1^3 + 3^3 + 5^3 + 7^3 + 9^3 + 11^3 + 13^3 + 15^3$.

(c) *Friendly (amicable) numbers*: Two numbers are called *friendly* or *amicable numbers* if the *proper* divisors of one add up to the other number. Greeks found that 220 and 284 are friendly numbers. We may see that the proper divisors of 220 are 1, 2, 4, 5, 10, 11, 20, 22, 44, 55 and 110 and all these add up to 284; similarly the proper divisors of 284 are 1, 2, 4, 71 and 142 and all these add up to 220. Pythagoreans defined friendship through these two numbers by declaring a friend is "one who is the other I such as 220 and 284". In 1636, Fermat obtained the second set of friendly numbers as 17,296 and 18,416. Soon after, Descartes obtained another set 9,363,584 and 9,437,056. In 1750, Euler listed 60 such pairs. But all these greats missed a pair 1,184 and 1,210 reported by 16-year old Paganini in 1866. Historians of mathematics found that Arab mathematician, Thabit ibn Qurra (836–901) mentioned the following method of obtaining friendly numbers by trial and error. He wrote that if three prime numbers a, b and c of the following form

$$\left.\begin{array}{l} a = 3 \times 2^x - 1 \\ b = 3 \times 2^{x-1} - 1 \\ c = 9 \times 2^{2x-1} - 1 \end{array}\right\} \tag{2.24}$$

with $x > 1$ can be found, then the numbers $2^x ab$ and $2^x c$ will be friendly numbers. But nothing is mentioned about how to choose the value of x. The reader may verify that by taking $x = 2, 4$ and 7, respectively one gets the friendly numbers found by the Greeks, Fermat and Descartes.

2.7 Taxicab numbers

There is a story behind this peculiar name for a class of numbers. Ramanujan fell seriously ill during his stay at Cambridge. Once when he was in a hospital in London, his mentor the famous English mathematician Godfrey Harold Hardy (1877–1947) came to visit him. Hardy told Ramanujan that he came to the hospital from the railway station by a taxi bearing an uninteresting number 1729.

Ramanujan immediately replied that this is, in fact, a very special number. This is the smallest number that can be expressed as a sum of cubes of two positive numbers in two different ways, viz. $10^3 + 9^3 = 12^3 + 1^3 = 1729$. There is no smaller number that can be written in this manner. Hardy was surprised to say the least. No wonder that John Edensor Littlewood (1885–1977), Hardy's collaborator and another great Cambridge mathematician, once remarked that all natural numbers are personal friends of Ramanujan. From that time 1729 is called the *Taxicab number*. Later on mathematicians searched for the smallest number which could be expressed as a sum of two cubes in three different ways. Now it is known that this number is $87{,}539{,}319 = 167^3 + 436^3 = 228^3 + 423^3 = 255^3 + 414^3$.

In fact this number is called the *Taxicab number* of order 3 and Ramanujan's number as the Taxicab number of order 2. The generalized Taxicab number of *n*th order, $T(n)$, is defined as the smallest number that can be expressed as a sum two positive cubes in *n* different ways. It has been proved that generalized Taxicab numbers exist for all values of *n*. Using supercomputers, Taxicab numbers of orders up to 6 have been obtained.

2.8 Pythagorean triples

If three integers X, Y and Z are related by the equation

$$X^2 + Y^2 = Z^2 \qquad (2.25)$$

then these three integers are called *Pythagorean triples*. It is evident that a right-angled triangle, satisfying Pythagoras's theorem can be drawn with sides of length X, Y and Z with the hypotenuse of length Z. Mathematicians starting from the Greek civilization have studied these numbers. Here we give a short discussion on Pythagorean triples. If the triple have no common factor between them, i.e. if the triple are co-primes, then such a triple is called a *primitive Pythagorean triple*. We will denote the primitive ones by small letters, like x, y and z. It is needless to say that from a primitive set of triples we can generate an infinite set of Pythagorean triples X, Y and Z by writing

$$X = kx, \ Y = ky \text{ and } Z = kz, \qquad (2.26)$$

with k = any positive integer.

Primitive Pythagorean triples can be generated by using the following Euclid's formula:

$$x = m^2 - n^2, \ y = 2mn \text{ and } z = m^2 + n^2 \qquad (2.27)$$

where the co-prime positive integers m and n satisfy the following conditions: (i) $m > n$ and (ii) $(m - n)$ is odd.

The reader is advised to generate some primitive Pythagorean triples by taking different values of m and n satisfying these two prescribed conditions. For example, $(m, n) = (2, 1), (3, 2), (4, 1), \ldots$, etc. You are asked also to notice

(a) Exactly one of x and y is a multiple of 3
(b) Exactly one of x and y is a multiple of 4
(c) Exactly one of x, y and z is a multiple of 5
(d) The product of x and y is a multiple of 12
(e) The product of three primitive triples is a multiple of 60.

An infinite set of Pythagorean triples exist where the smaller two differ by 1. For example, $3^2 + 4^2 = 5^2$, $20^2 + 21^2 = 29^2$.

All such triples are, in fact, given by the form

$$\left(\frac{a-1}{2}\right)^2 + \left(\frac{a+1}{2}\right)^2 = z^2 \tag{2.28}$$

where the integers a and z are the roots of the equation

$$a^2 - 2z^2 = -1 \tag{2.29}$$

For the solution of eq. (2.29) see Appendix A2.

An infinite set of primitive Pythagorean triples also exist where the larger two differ by 1. For example, $3^2 + 4^2 = 5^2$, $5^2 + 12^2 = 13^2$.

All such triples can be written as

$$(2m+1)^2 + (2m^2 + 2m)^2 = (2m^2 + 2m + 1)^2 \tag{2.30}$$

where m is any positive integer. Note that only the smallest primitive triple (3, 4 and 5) consist of three consecutive integers. Interestingly, the right angled triangle with sides measuring 3, 4 and 5 has an area 6 (i.e. the next integer) units. Also recall Section 2.3.3(d).

2.9 Sequences

In this section, we discuss in brief a very famous sequence of positive integers around which a lot of mathematics has been developed and is still continuing. This sequence is known in the western world by the name *Fibonacci sequence*. Fibonacci is the common name by which the Italian mathematician Leonardo of Pisa (c. 1170–c. 1250) is known all over the world. He was the most talented mathematician of his era in Europe. In 1202 he wrote a famous book *Liber Abaci*, where he introduced the modern Hindu-Arabic numeral to the western world. The golden age European Mathematics started after that. In this book, he introduced a famous sequence representing a crude model of the growth of rabbit population in his garden. Fibonacci sequence is written as 1, 1, 2, 3, 5, 8, 13, 21, 34, 55, 89, 134, ... The first two terms are written as 1 and 1 and then onward every subsequent number is the sum of the immediately two previous numbers. Mathematically we write, the nth term in the sequence as follows:

$$F_n = F_{n-1} + F_{n-2} \text{ for } n \geq 3 \tag{2.31}$$

with $F_1 = 1$ and $F_2 = 1$. The numbers appearing in this sequence are called *Fibonacci numbers*. Frequent appearance of these numbers in various natural objects like plants, flowers, etc. and their connections with other mathematical entities will be discussed in some details in the next chapter on real numbers.

Recently it has been brought to the attention of the world that at least 50 years before Fibonacci, great Indian poet and linguist Hemachandra discussed these numbers in an altogether different context. Hemachandra introduced these numbers to describe the different possible rhythms with a specified number of beats. Following Hemachandra, the following table of total number of beats and corresponding total number of possible rhythms can be constructed:

Total number of beats	1	2	3	4	5	6	7	8	...
Total number of rhythms	1	2	3	5	8	13	21	34	...

The appearance of Fibonacci numbers in the total number of rhythms can be easily seen with the only difference of the absence of two 1's in the beginning. Thus, Hemachandra sequence can be mathematically written as

$$H_n = H_{n-1} + H_{n-2} \text{ for } n \geq 3 \qquad (2.32)$$

with $H_1 = 1$ and $H_2 = 2$.

The two sequences can be easily related through

$$F_{n+1} = H_n \text{ for } n \geq 1 \qquad (2.33)$$

with $F_1 = 1$. Today this sequence is called *Fibonacci-Hemachandra sequence*.

The approach of generating Fibonacci-Hemchandra type sequence is useful for solving various arithmetic problems. One example follows.

Suppose you have to climb a staircase having 6 stairs. You can climb 1, 2 or 3 stairs at a time, i.e. you may not skip any stair or may skip 1 or even 2 stairs by jumping. How many different ways can you climb the staircase? Here we note that the first three stairs we can climb as follows.

First stair: only 1 way, i.e. $(0 + 1)$, we denote $N1 = 1$.

Second stair: 2 ways, one at a time or directly jumping on to this level, i.e. $(1+1)$ and $(0 + 2)$, we denote $N2 = 4$.

Third stair: 4 ways $(1 + 1 + 1)$, $(1 + 2)$, $(2 + 1)$, $(0 + 3)$, we denote $N3 = 4$.

Notice that you can reach the 4th stair from the first (skipping 2 stairs), the second (skipping 1 stair) and the third (without skipping any stair). We write $(1 + 3)$, $(1 + 1 + 2)$, $(0 + 2 + 2)$, $(1 + 1 + 1 + 1)$, $(1 + 2 + 1)$, $(2 + 1 + 1)$, $(0 + 3 + 1)$. Notice we have added 3 to $N1$, 2 to each of $N2$'s and 1 to each of $N3$'s. Thus, we get $N_4 = N_1 + N_2 + N_3 = 1 + 2 + 4 = 7$.

Similarly, $N_5 = N_2 + N_3 + N_4 = 2 + 4 + 7 = 13$ and $N_6 = N_3 + N_4 + N_5 = 4 + 7 + 13 = 24$.

If you could climb only 1 or 2 (by skipping one) stairs at a time, then the answer would have been given by Hemchandra number $H_6 = 13$. This is left as an exercise for the reader.

2.10 Miscellany

In this last section on integers we will discuss a couple of important concepts and curious information.

2.10.1 Representation of large numbers

Very often we have to deal with large numbers and to write them in decimal notation is tedious. Also a lot of space will be needed.

Therefore, some other notations are common and you must be already exposed to it. One such notation is the factorial notation. We write the product of all natural numbers 1, 2, 3, ... up to n as $n!$. This is quite compact as, 50! if written in decimal notation, will consist of 65 digits.

Another common way is the use of exponential notation, which is essential to carry out computation in different areas in mathematics. One subtle point may be worthwhile to point out while using multiple exponents. It must be remembered that

$$(a^m)^n = a^{mn} \tag{2.34}$$

$$\text{while } a^{m^n} = a^{(m^n)} \tag{2.35}$$

and the above two expressions are very different. This can be easily appreciated by calculating

$$2^{2^{2^2}} = 2^{2^4} = 2^{16} = 65{,}536$$

Multiple exponentiations are right-associative processes, carried out from right to left. The use of parenthesis is necessary to indicate if left-association is needed, which will result in a different final result. For example,

$$\left(\left((2)^2\right)^2\right)^2 = \left((4)^2\right)^2 = (16)^2 = 256$$

With multiple exponents the result increases very rapidly. Gauss called the *number* 9^{9^9} *measurable infinity*! This number has $10^{369,693,100}$ digits. There is not enough space in the universe to write this number using this font size with an ink of even one molecule thickness. Mathematicians have calculated the last ten digits of this number as ... 1,045,865,289. Recently mathematicians have created some more notations (like up-arrow towers) as even exponential notation was not suitable to express the numbers involved.

2.10.2   Fascination of integers

Mathematicians of different era are fascinated with curious properties of various integers. In fact, there is enormous material in print and in the web discussing unexpected and interesting information about prime and composite numbers. Either an interesting property has been noticed or proved about unsuspecting integers apparently without any special characteristic. Here we mention only two such examples.

Fermat noticed that the number 26 is sandwiched between a perfect square (= 25) and a perfect cube (= 27). He challenged English mathematicians John Wallis (1616–1703) and Kenelm Digby (1603–65) to find one more such number. When they failed, Fermat announced there is no other such number. Euler proved this assertion and the proof was perfected later on.

Henri Brocard (1845–1922) and later on Ramanujan independently noticed that $(n! + 1)$ is a perfect square only for $n = 4, 5$ and 7 as written below

$$4! + 1 = 25 = 5^2$$
$$5! + 1 = 121 = 11^2$$
$$7! + 1 = 5041 = 71^2$$

The question is whether 7 is the highest value of n for which $(n! + 1)$ is a perfect square? No other value of n has been found by searching up to 109. But the non-existence has not been proved.

Chapter 3

Real Numbers

3.1 Introduction

In the last chapter we have seen the number line on which the integers are represented by discrete points with equal gap between two consecutive integers. The concept of number is then extended to fill up all these gaps with what is called *real number continuum*. In other words, all the points on the number line now represent numbers which are called *real numbers*. Thus, integers form a subset of the real numbers. The number line is now referred to as the real number line. In this chapter, we define, classify and discuss different representations of real numbers. Furthermore, besides discussing some curious patterns, we also point out surprising relationships between some numbers. Then we take up a few special real numbers. These special numbers are important universal constants which are frequently used in various branches of mathematics. Finally, it is shown how some special real numbers are used to describe various proportions and forms found in the self-made tapestry of nature.

3.2 Rational numbers

We already mentioned that by dividing an integer by another we do not always get another integer. So we create another class of number, called *rational number*. A rational number R is defined using two integers, say p and q as

$$R = \frac{p}{q} \text{ with } q \neq 0 \tag{3.1}$$

Integers are included in the category of rational numbers with $q = 1$. Unless otherwise stated, we assume that p and q are co-primes, implying there is no common divisor (except trivial 1) between p and q. This fact can also be said as "the rational number has been expressed in its simplest or lowest form". With $q > p$, we call the number a rational fraction, R_f, with

$$R_f = \frac{p}{q} \text{ with } q > p \tag{3.2}$$

The integer p is called the *numerator* and q the *denominator* of the fraction. A rational number R greater than 1 has an integer part I and a fractional part $F(= R_f)$, i.e.

$$R = I + F \tag{3.3}$$

The segment of the real number line between 0 and 1 cannot be filled up by only rational fractions. Similarly, any segment of the real number line cannot be filled up by only rational numbers. The gaps left in the real number line, after plotting all the infinite rational numbers, are filled up by another infinite set called *irrational numbers*.

3.3 Irrational numbers

Greek geometers gave the identity of a number to any length which can be drawn geometrically with an arbitrarily assumed unit length. They could easily draw squares and regular pentagons of unit side and hence their diagonals. To their utter surprise they found that the lengths of these diagonals could not be expressed as rational numbers. For example, we can calculate the length of the diagonals of a unit square as $\sqrt{2}$ and that of a unit regular pentagon as $\frac{\sqrt{5}+1}{2}$. Following Euclid we prove below that $\sqrt{2}$ cannot be a rational number. If $\sqrt{2}$ is a rational number, then we write it in the lowest form as

$$\frac{p}{q} = \sqrt{2} \text{ or, } \frac{p^2}{q^2} = 2 \text{ or, } p^2 = 2q^2 \tag{3.4}$$

So p^2 and hence p is an even number. Let $p = 2m$, where m is an integer. From eq. (3.4), we get

$$q^2 = 2m^2 \tag{3.5}$$

Hence q^2 and so q is also an even number. Consequently we land in a contradiction as both p and q cannot be even (having a common divisor 2), since we assumed that they do not have any common divisor (other than 1).

Therefore, the conclusion is $\sqrt{2}$ cannot be a rational number. Thus, another class of numbers, irrational numbers, was identified to fill up the real number line. Greek mathematician Eudoxus gave a geometric theory of irrational numbers and in the second half of the nineteenth century an arithmetical theory of irrational numbers was developed. Irrational roots of rational numbers, like $\sqrt{2}$, $\sqrt[3]{5/2}$, etc. are sometimes called *surds*. It must be emphasized that it has been proved that all irrational numbers, e.g. $\sqrt[3]{2}$ are not geometrically, i.e. using only a compass and an unmarked straight edge, constructible.

3.4   Transcendental numbers

If a number can be obtained as a root of any polynomial equation, then such numbers are called *algebraic numbers*. Obviously all rational numbers p/q are roots of linear equations of the form $qx - p = 0$, and hence belong to the category of algebraic numbers. But irrational numbers can be both algebraic and non-algebraic. Such non-algebraic irrational numbers are called *transcendental numbers*. It can be easily seen that that irrational numbers, like $\sqrt{2}$, $\sqrt[3]{5/2}$ and $\frac{\sqrt{5}+1}{2}$, respectively, satisfies the following algebraic equations

$$x^2 - 2 = 0,$$
$$2x^3 - 5 = 0,$$
$$\text{and } x^2 - x - 1 = 0$$

and are hence algebraic. But mathematicians could write some numbers which were proved as non-algebraic; that means it was proved that those numbers could not be roots of any polynomial equation. Later on, some famous universal mathematical constants like π, e, e^π, etc. were also proved to be transcendental numbers. To prove that a number is irrational or transcendental is enormously difficult. For example, till today nobody knows whether π^e is transcendental or not. A general theorem states that for any algebraic number $a \neq 0, 1$ and any algebraic irrational number b, the number a^b is transcendental. This was a famous theorem proved in the last century. An immediate conclusion is that the number $2^{\sqrt{2}}$, called *Hilbert's number*, is transcendental.

3.5   Decimal and binary representation

We might recall that in chapter 2 it was explained how integers are represented in decimal notation by a series of non-negative integral powers of 10 using 10 digits, viz. $0, 1, 2, \ldots, 9$. In a similar manner, using all these digits rational fractions are expressed by a series of negative integral powers of 10. We write a rational fraction R_f, with $0 < R_f < 1$ as

$$R_f = \sum_{k=1}^{n} \frac{b_k}{10^k} \tag{3.6}$$

where b_k denotes one of the digits $0, 1, 2, \ldots, 9$ with the restriction $b_n \neq 0$. Using a decimal point we ultimately write

$$R_f = 0.b_1 b_2 b_3 \ldots b_n \tag{3.7}$$

A rational number greater than 1 is written by replacing the zero before the decimal point in (3.7) by its integer part I written using (2.2).

It may be mentioned that for some fractions the value of n may be limited whereas for some others the digits after the decimal point keep on repeating endlessly in a periodic fashion. For example, let us look at the following decimal expressions of some fractions:

$$\frac{3}{20} = 0.15; \quad \frac{1}{3} = 0.333\ldots333\cdots = 0.\dot{3} \text{ or, } 0.\overline{3}$$

$$\frac{7}{6} = 1.1666\ldots666\cdots = 1.1\dot{6} \text{ or, } 1.1\overline{6}$$

$$\frac{1}{7} = 0.142857142857\ldots142857\cdots = 0.\dot{1}4285\dot{7} = 0.\overline{142857}$$

(3.8)

It can be easily seen that in the first example above, the decimal expression terminates, whereas in the others the decimal expression continues with periodic repetitions of a digit or a string of digits. In the terminating decimal, by adding a 0 at the end we can rephrase that the digit 0 repeats indefinitely. Such periodic repetition of digits in the decimal expression is the characteristics of rational numbers. (For more mathematically curious readers, a simple proof of this statement is given at the end of this section). However, we do not write this repeating zero at the end of the decimal expression of a fraction as we also do not write zeros at the beginning of the decimal expression of an integer. In such a situation, it is called a *terminating decimal*, whereas the others with a periodic tail are called *non-terminating* or *recurring decimals*. Recurring decimals are written by writing the periodic pattern of digits only once. Either two dots are put on the first and the last digits or a bar is drawn over the periodic pattern. The condition for a terminating decimal is that the denominator of the fraction must be expressible in the form $2^{n_1} \times 5^{n_2}$, where n_1 and n_2 are non-negative integers.

Before discussing the decimal representation of irrational numbers, we show here that $0.\overline{9} = 1$, not approximately but exactly equal to 1. To see this we put

$$x = 0.\overline{9} = .999999\ldots$$

Multiplying both sides by 10 we write $10x = 9.9999\ldots$ Now subtracting the above equation from this, one gets $9x = 9$ or, $x = 1$.

Irrational numbers are characterized by non-terminating, non-recurring appearance of digits indefinitely after the decimal points. Thus, irrational numbers can never be explicitly written in decimal notation. These numbers can only be approximated by computers. For example, we give below some approximate decimal representation of a few well known irrational numbers.

$$\pi = 3.1415926535897\ldots$$

$$e = 2.7182818284590452\ldots$$

$$\sqrt{2} = 1.41421356237\ldots$$

Some of these have been calculated to a trillion, billion or million places after the decimal point. No repeating pattern could be observed. The first two numbers mentioned above have been proved to be transcendental. Initially and also later some transcendental numbers were constructed and proved to be so. Two such examples are given below:

(i) *Liouville's number*: 0.11000100000000000000000000100 . . .

This number consists of a string of 1's and 0's after the decimal point. The digit 1 appears at all n!th places, where n indicates any natural number. For example, at 1!(= 1), 2!(= 2), 3!(= 6), 4!(= 24)th places.

(ii) *Champernowne's number*: 0.12345678910111213141516 17 . . .

Here the decimal point is followed by writing all the natural numbers sequentially.

Just like integers, rational fractions can also be expressed in binary notation (base −2), using the power series of 2 with negative integral powers. So we can write [Recall eq. (2.3)] any rational fraction. R_f in binary notation as

$$R_f = \sum_{k=1}^{n} \frac{b_k}{2^k} \tag{3.9}$$

where b_k denotes one of the digits 0 or 1 with the restriction $b_n \neq 0$.

Thus, we write

$$R_f = 0.b_1 b_2 b_3 \ldots b_n \tag{3.10}$$

For any rational number greater than 1, one has to replace the zero in front of the decimal point by the binary expression of the integer part as given by (2.3); so finally we get a string of 0's and 1's with a decimal point separating the integer and fraction parts.

(As stated earlier, here we give the proof of the statement that the digits in the decimal expression of any rational number must show a periodic tail. We start with the derivation of a famous rational approximation of the universal constant, $\pi \approx \frac{355}{113}$.

The reader is advised to obtain the decimal expression of the rational fraction $(355/113) = 3 + (16/113) = 3.141592920353982300884 \ldots$

At this stage you may be fed up that still no periodicity is observed! But if you study the process of obtaining this value you should notice that every stage while dividing by 113, a new digit is appearing without exhibiting periodicity. This is because at every stage a new remainder is obtained during the division by 113. But once the value of the remainder is repeated, i.e. one of the previously obtained values, then the periodic tail of digits in the decimal expansion must be exhibited. Now this repetition of the remainder is ensured. This follows from the fact that the remainder has to be a number in the range 0 to 112, i.e. there can be at most 113 different remainders and not infinitely many. With the remainder 0, the fraction is a terminating decimal. With any other remainder, it has a periodic tail. Since at most 112 different non-zero remainders are possible, the non-zero remainder must repeat somewhere definitely within 112 steps after a non-zero digit appears for the first time. Mathematicians call it *pigeonhole principle,* which simply states if there are n pigeonholes and m pigeons with $m > n$, then if all the pigeons are put in the pigeonholes, at least one hole must contain more than one pigeons. Here the remainders are the pigeons and the maximum number of possible remainders is the number of pigeonholes. The finite number of number of remainders (pigeonholes) ensures identical remainders and repetition of the finite periodic chain of digits in the decimal expression. *Pigeonhole principle* is a very useful mathematical principle which we use to solve different types of problems in chapter 5).

3.6   Continued fraction representation

In chapter 1 we encountered an infinite continued fraction given by Ramanujan to solve the puzzle given in *The Strand* magazine. All rational numbers can always be expressed by a finite continued fraction with only 1 as the numerator as given below. Such a continued fraction with only 1 as numerator is called a *simple continued fraction* and written as

$$R = A_0 + \cfrac{1}{A_1 + \cfrac{1}{A_2 + \cfrac{1}{A_3 + \cfrac{\ddots}{+\cfrac{1}{A_n}}}}} \tag{3.11}$$

where A_0 represents the integer part and all other A_i's are integers. To express a continued fraction in a compact form, occupying less space, we write

$$R = (A_0; A_1, A_2, A_3, \ldots, A_n) \tag{3.12}$$

As an example, let us write the rational fraction 228/611 as a simple continued fraction. First note that the integer part $A_0 = 0$. Then write

$$\frac{228}{611} = \cfrac{1}{\frac{611}{228}} = \cfrac{1}{2 + \frac{155}{228}} = \cfrac{1}{2 + \cfrac{1}{\frac{228}{155}}} = \cfrac{1}{2 + \cfrac{1}{1 + \frac{73}{155}}} = \cfrac{1}{2 + \cfrac{1}{1 + \cfrac{1}{\frac{155}{73}}}}$$

$$= \cfrac{1}{2 + \cfrac{1}{1 + \cfrac{1}{2 + \frac{9}{73}}}} = \cfrac{1}{2 + \cfrac{1}{1 + \cfrac{1}{2 + \cfrac{1}{\frac{73}{9}}}}} = \cfrac{1}{2 + \cfrac{1}{1 + \cfrac{1}{2 + \cfrac{1}{8 + \frac{1}{9}}}}}$$

Thus, this continued fraction in compact form is written as $(0; 2, 1, 2, 8, 9)$.

Irrational numbers can be written as a continued fraction which never ends. But one thing is worth mentioning that in the decimal notation no periodic pattern is found for irrational numbers. But it has been proved that for quadratic irrationals like, $\sqrt{2}, \sqrt{3}, \frac{\sqrt{5}-1}{2}$, etc. the continued fraction representation exhibits a periodic repetition. For example, we can show that

$$\sqrt{2} = (1; 2, 2, 2, \ldots), \quad \sqrt{3} = (1; 1, 2, 1, 2, 1, 2, \ldots), \quad \frac{\sqrt{5}-1}{2} = (0; 1, 1, 1, \ldots)$$

and so on.

Now we show that the continued fraction $(1, 2, 1, 2, 1, 2, \ldots) = \sqrt{3} - 1$. Towards this end, we first write

$$x = \sqrt{3} - 1 = \cfrac{1}{1 + \cfrac{1}{2 + \cfrac{1}{1 + \ddots}}} = \cfrac{1}{1 + \cfrac{1}{2+x}} \text{ or, } x = \frac{2+x}{3+x} \text{ or, } x^2 + 2x - 2 = 0$$

Solving and discarding the negative infeasible root, we get $x = \sqrt{3} - 1$. It may be mentioned that $\frac{\sqrt{5}-1}{2}$, expressed by using only the digit 1 in the continued fraction form, is called the *worst irrational number*. This number is again discussed in Section 3.9.1.

It must be emphasized that in the procedure mentioned above we replaced an infinite continued fraction (chain) repeatedly by x to solve for it. It is worthwhile to caution against blind application of this technique. One needs to be cautious and prove that the infinite continued fraction does converge. Without getting into the discussion of the convergence proof, we discuss a paradox created by a blind application of this technique.

Suppose we want to solve for x in the following equation where the exponentiation continues forever:

$$x^{x^{x^{x^{x^{\cdot^{\cdot^{\cdot}}}}}}} = 2 \tag{3.13}$$

Replacing the infinite chain by 2 we can as well write $x^2 = 2$, or, $x = \sqrt{2}$. Now if we solve eq. (3.13) with the righthand side equal to 4, then by applying the same technique we get $x^4 = 4$, when again $x = \sqrt{2}$. Then obviously the question arises what is the correct value of the (infinite) expression

$$\sqrt{2}^{\sqrt{2}^{\sqrt{2}^{\cdot^{\cdot^{\cdot}}}}}$$

2 or 4 ? The answer to this riddle is given later after we discuss a bit more on the special number e in Section 3.9.3.

3.7 Different types of mean of a set of real numbers

A set of real numbers very often represent data regarding various parameters involved in a phenomena. In such cases we are interested in different kinds of average values of such a data set. Depending on the parameter of the data set, different kinds of average values reflect the average behaviour. Some data set may consist of only positive values, whereas in some others the set may have both positive and negative values (like voltages in a circuit). Some simple relationships exist between these different kinds of averages or mean values. These relationships are also found to be very useful in solving some problems involving numbers. In this section, we define these various means and write the useful relationships (without the general proof) between these means.

Let a data set be represented by real numbers $(x_1, x_2, x_3, \ldots, x_n)$. Then the root mean square (RMS) value is defined as

$$\text{RMS} = \sqrt{\frac{\sum_{i=1}^{n} x_i^2}{n}} \tag{3.14}$$

It means that we follow the reverse order of the word RMS, i.e. first take the Squares, then the Mean and then the square Root. This is most commonly used for data set having both positive and negative values. You may have heard that the RMS value of the voltage is 220V, which gives the average measure of the magnitude.

The arithmetic mean (AM) is defined as

$$\text{AM} = \frac{\sum_{i=1}^{n} x_i}{n} \tag{3.15}$$

It means that we take the sum and then divide by the number of data points; like the average weight or height of all the students in a class. No data point in that case is negative.

The geometric mean (GM) is defined as

$$\text{GM} = \sqrt[n]{\prod_{i=1}^{n} x_i} \tag{3.16}$$

It means that we take the product and then obtain the nth root, where n is the number of data points.

The harmonic mean (HM) is defined as

$$\frac{1}{\text{HM}} = \frac{1}{n} \left(\sum_{i=1}^{\infty} \frac{1}{x_i} \right) \text{ or, HM} = \frac{n}{\sum_{i=1}^{n} \frac{1}{x_i}} \tag{3.17}$$

It means that we first take the arithmetic mean of the reciprocals of the values and then take the reciprocal of that.

It can be proved that for a set of positive real numbers

$$\text{RMS} \geq \text{AM} \geq \text{GM} \geq \text{HM} \tag{3.18}$$

The equality sign in (3.18) holds when all the data values are same and all the means are also equal to that unique value of the data set. The relation (3.18) is very useful in solving problems involving numbers. The proof of (3.18) for just two values of data say x_1 and x_2 can be done very easily. The reader is encouraged to try this proof. The general proof starts from this step and then the method of induction is used. In what follows we give the geometric constructions of these four means for only two values of data. Refer to Figs 3.1(a) and (b); in these figures OA = x_1 and OB = x_2, and the diameter and the centre of the semicircle are AB and C, respectively. In Fig. 3.1(a), OD is the RMS, CD is the AM and OE is the GM. In Fig. 3.1(b) EF is the HM. The proof of this geometric construction is also left as an exercise for the reader. From these constructions, the relationships (3.18) are clearly established for two values of data and the equality sign holds when the radius of the semicircle represents all the four means and also the two values of data.

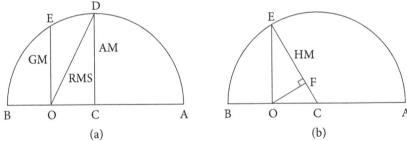

Fig. 3.1: Geometric construction of different types of mean of two positive real numbers

3.8   Special rational numbers

In this section, we discuss some special rational numbers, either defined in a particular way or for some special purposes. Sometimes there may be some curious properties of these special rational numbers.

3.8.1   Bernoulli numbers

We know the sum of the first n consecutive numbers. We now discuss how to obtain the sum of the integral powers of the first n consecutive natural numbers. These formulas (discussed in the Appendix B) were first obtained by Johann Faulhaber and were included in the famous text of Jacob Bernoulli. These formulas use certain rational numbers which were called *Bernoulli numbers* by Euler. Then onwards, these special rational numbers are called *Bernoulli numbers*. Before proceeding any further, let us reproduce the formula for the sum of the integral powers of the first n consecutive natural numbers as

$$S^{(k-1)} = 1^{k-1} + 2^{k-1} + 3^{k-1} + \cdots + n^{k-1}$$

$$= \frac{1}{k} \left[B_0 n^k + B_1 \binom{k}{1} n^{k-1} + B_2 \binom{k}{2} n^{k-2} + B_3 \binom{k}{3} n^{k-3} + \cdots \right] \quad (3.19)$$

where $k \geq 1$ is a natural number. The constants $B_0, B_1, B_2, B_3, \ldots$ are known as *Bernoulli numbers.* The first few Bernoulli numbers are obtained as

$$\left. \begin{array}{l} B_0 = 1, \quad B_1 = \frac{1}{2}, \quad B_2 = \frac{1}{6}, \quad B_3 = B_5 = B_7 = \cdots = 0 \\[2mm] B_4 = -\frac{1}{30}, \quad B_6 = \frac{1}{42}, \quad B_8 = -\frac{1}{30}, \quad B_{10} = \frac{5}{66}, \quad B_{12} = -\frac{691}{2730}, \quad \ldots \end{array} \right\} \quad (3.20)$$

It must be mentioned that obviously $B_0 = 1$ and it can be noticed in (3.20) that except B_1 all odd order Bernoulli numbers are zero. Moreover, even order Bernoulli numbers (starting from the second) are seen to be alternately positive and negative. Eventually very high order Bernoulli numbers involve rather large numbers.

These Bernoulli numbers have been widely used in various branches of mathematics. More and more Bernoulli numbers have been calculated by many famous people. Euler calculated the first 30, Ohm the first 62 and Adams up to the first 124. These numbers were so useful that Charles Babbage was requested to compute these numbers with his computing machine (the forerunner of modern day calculators and computers). The consecutive Bernoulli numbers can be recursively calculated with the knowledge of all the lower order Bernoulli numbers. This is done by calculating the following linear algebraic equations:

$$\left. \begin{array}{l} B_2 = B_2 - 2B_1 + 1 \text{ or, } B_1 = \frac{1}{2} \\[2mm] B_3 = B_3 - 3B_2 + 3B_1 - 1 \text{ or, } B_2 = \frac{1}{6} \\[2mm] B_4 = B_4 - 4B_3 + 6B_2 - 4B_1 + 1 \text{ or, } B_3 = 0 \\[2mm] B_5 = B_5 - 5B_4 + 10B_3 - 10B_2 + 5B_1 - 1 \text{ or, } B_4 = -\frac{1}{30} \end{array} \right\} \quad (3.21)$$

The righthand sides of eq. (3.21) are obtained from the Pascal's triangle discussed in Section 2.3.4 (Fig. 2.2) by starting from the third row and putting alternately + and − signs. This process can go on for all orders. Obviously B_i can be calculated if all the Bernoulli numbers up to B_{i-1} are known using the previous equations.

3.8.2 Unit fractions

Egyptians used only reciprocals of natural numbers (greater than 1) as fractions. The numerator of an Egyptian fraction is always 1. These are also called *unit fractions.* All other fractions were written as sums of unit fractions like, $\frac{2}{3} = \frac{1}{2} + \frac{1}{6}$; but this breaking up in terms of unit fractions is not unique, as we can also write $\frac{2}{3} = \frac{1}{3} + \frac{1}{3}$. The only reason for discussing these not so convenient unit fractions is that there are some curious properties of certain unit fractions. First we consider those generated by the reciprocals of prime numbers. Let us consider the number of digits in the recurring periodic chain of the decimal expression of the unit fraction $\frac{1}{p}$, with p denoting a prime number. It will be found that $(p - 1)$ will always be divisible by this number (defining the chain length). Towards this end, consider the decimal expressions for $\frac{1}{3}$ and $\frac{1}{7}$, given in (3.8), i.e. the unit fractions generated by the reciprocals of prime numbers 3 and 7. In the first case $(3 - 1)$ is divisible by the chain length 1, and in the second case $(7 - 1)$ is divisible by the chain length 6. Of course, this statement will be true for reciprocals of all primes. The reader is advised to check this statement for unit fractions generated by the reciprocals of prime numbers 17 and 19. Now we consider the following equation involving three unit fractions:

$$\frac{1}{n} = \frac{1}{i} - \frac{1}{j} \tag{3.22}$$

If the number n is a prime number p, then there exist only a unique pair of values for (i, j); whereas if the number n is a composite number C, then the number of pairs of values for (i, j) is always more than one. In fact, the number of possible pairs is given by the total number of those divisors of n^2, which are smaller than n. In chapter 2, we have already noted that p^2 has only three divisors, viz. 1, p and p^2; thus 1 is the only divisor smaller than p. It can be easily seen that for all values of $n > 2$, one possible value of $i = n - 1$, when $\frac{1}{j} = \frac{1}{n-1} - \frac{1}{n} = \frac{1}{n(n-1)}$. With $n = p$, the unique values of i and j are given by $(n - 1)$ and n, $(n - 1)$, respectively. For example, we may note that $\frac{1}{3} = \frac{1}{2} - \frac{1}{6}$; there is no other way we can express the reciprocal of 3 as the difference between two unit fractions. But the reciprocal of 4, we can write $\frac{1}{4} = \frac{1}{3} - \frac{1}{12}$ and also $\frac{1}{4} = \frac{1}{2} - \frac{1}{4}$. The reader is advised to find the possible values of (i, j) with $n = 7$ and $n = 24$ to verify that for the prime number 7, there is a unique pair of (i, j) and for the composite number 24 the number of such pairs is 10 (since 1, 2, 3, 4, 6, 8, 9, 12, 16, 18 all these 10 divisors of 576 are less than 24). A hint to solve this exercise may be obtained by carefully studying the example with $n = 4$, i.e. obtain all the divisors of 16 which are less than 4 and observe the values of i. For a general proof of the statement made above, and how to find all the values of (i, j) for a given n, see Appendix C.

Sometimes curious periodic patterns are generated by the decimal representations of some unit fractions. In eq. (3.8) we have seen that the decimal representation of certain rational fractions exhibit a periodic repetitions of a chain of digits immediately following the decimal point; for example $\frac{1}{7} = 0.\overline{142857}$. In general for such a rational fraction, $R_f = 0.\overline{a_1 a_2 a_3 \ldots a_n}$. From the decimal expression, the corresponding rational fraction can be retrieved by writing $R_f = \frac{a_1 a_2 a_3 \ldots a_n}{999 \ldots 9}$, where the denominator contains n numbers of 9. You may verify $\frac{142857}{999999} = \frac{1}{7}$. It may also be pointed out that $\frac{1}{27} = 0.\overline{037}$ and $\frac{1}{37} = 0.\overline{027}$, i.e. as if only the digits 2 and 3 have been interchanged!

Now we discuss a class of rational fractions whose decimal representations show curious periodic patterns immediately following the decimal point. Towards this end, first let us consider the decimal representation of $\frac{1}{81}$.

$$\frac{1}{81} = \frac{1}{9^2} = 0.\overline{012345679}$$

We should note that all the digits 0 to 9, except 8, appear sequentially following the decimal point. This chain of digits then repeats periodically. Next let us consider the fraction $\frac{1}{9801}$.

$$\frac{1}{9801} = \frac{1}{99^2} = 0.\overline{000102030405060708091011 12\ldots 969799}$$

Note that all the two digit numbers, i.e. 00, 01, 02, ,,, 97, 99 (except 98) appear sequentially and then this chain of numbers repeats periodically.

Similarly, the decimal representation of the fraction $\frac{1}{998001}$ is

$$\frac{1}{998001} = \frac{1}{999^2}$$
$$= 0.\overline{000001002003\ldots 009010011012\ldots 099100101102\ldots 996997999}$$

Note that all the three digit numbers, like, 000, 001, 002, 003, …, 009, 010, 011, …, 099, 100, 101, 102, …, 996, 997, 999 (the missing number, as expected, is 998) appear sequentially and then this long chain of numbers repeats periodically. The simple mathematics behind these curious decimal representations is explained in the Appendix D.

As the last example of a curious decimal expression of a unit fraction, we consider $\frac{1}{89} - 0.01123595505\ldots$ Apparently no pattern is observed in this decimal expression. But it has been proved that this number can be expressed as the sum of the numbers written below:

$$0.01123595505 = 0.01 + 0.001 + 0.0002 + 0.00003 + 0.000005 + 0.0000008$$
$$+ 0.00000013 + \ldots$$

We may observe that on the righthand side if we consider only the significant digits in each term, we get the sequence 1, 1, 2, 3, 5, 8, 13, …, i.e. the Fibonacci sequence defined in Section 2.9.

At this stage it may not be out of place to mention that just like Pascal's triangle (Section 2.3.4) was made using integers, an interesting pattern, shown in Fig. 3.2, called *Leibniz's harmonic triangle* can be constructed using the unit fractions.

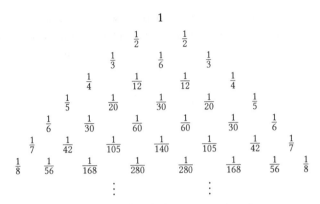

Fig. 3.2: Leibniz's harmonic triangle

The method of construction of Leibniz's harmonic triangle is similar to that of Pascal's triangle. We start with $1 (= 1/1)$ at the top and then write k numbers of unit fractions in the kth row making a triangle at each stage. First eight rows are shown in Fig. 3.2. The element at both ends of the kth row is the unit fraction $(1/k)$. Then the intermediate elements are filled up as follows. Take the difference of the two fractions, one that you find by moving diagonally left to the immediate upper row and the other by moving horizontally left in the same row. Mathematically this rule can be described as the element of the mth row and nth column $L_{m,n}$ is given by

$$L_{m,n} = L_{m-1,n-1} - L_{m,n-1} \qquad (3.23)$$

Thus, the intermediate number in the third row is $(1/2) - (1/3) = 1/6$. The reader is advised to verify some more entries to be convinced about the procedure. It may be noticed that if we move down along the left edge of the triangle, we encounter reciprocals of the natural numbers. If we move similarly along the next parallel line we get reciprocals of twice the triangular numbers. Another exercise for the reader is to verify that (3.23) can also be written as

$$L_{m,n} = L_{m,1}/P_{m,n} \qquad (3.24)$$

where $P_{m,n}$ represents the element at the mth row and nth column in the Pascal's triangle.

3.9 Special irrational and transcendental numbers

As mentioned earlier, some real numbers are easily identified as irrational, but some defined from geometry or as a series may not be easy to identify as transcendental. In this section, we discuss four very special numbers, each one of which now has a universal symbol representing it. One of these and its reciprocal are clearly shown to be irrational but not transcendental (Golden Sections, ϕ and ψ), two very commonly used universal mathematical constants (π and e)

have been proved to be transcendental. The last one called *Euler's constant* (γ), when calculated up to more than a million places after the decimal point, exhibits non-periodicity. This number is believed to be irrational (may be even transcendental), but has not yet been proved to be so. Now we briefly discuss each of these separately because of their importance and frequent appearances not only in various branches of mathematics but also in other subjects like physics, chemistry and biology.

3.9.1 Golden sections

Greek geometers, artists and architects were mesmerized by two irrational numbers, viz. $\frac{\sqrt{5} \mp 1}{2}$, which are seen to be reciprocal of each other. These are called *golden sections* or *ratios*. For subsequent use in this section, we introduce the symbols

$$\phi = \frac{\sqrt{5} - 1}{2} = 0.618034\ldots$$

$$\text{and } \psi = \frac{\sqrt{5} + 1}{2} = 1.618034\ldots \tag{3.25}$$

These are among the earliest examples of irrational numbers. Greek geometers found that these numbers can be drawn geometrically and they define the ratio of the lengths of the side and the diagonal of a regular pentagon. Consequently a regular pentagon could be drawn geometrically, i.e. using an unmarked straight edge and a collapsible compass. Interestingly, both these numbers could be expressed using only the digit 1, one as an infinite continued fraction and the other as an infinite radical, as shown below:

$$\varphi = (0; 1, 1, 1, 1, \ldots)$$

$$\text{and } \psi = \sqrt{1 + \sqrt{1 + \sqrt{1 + \sqrt{1 + \ldots}}}} \tag{3.26}$$

Both these infinite expressions converge and the reader is advised to verify the values by the procedure explained in Section 3.6 for the continued fraction of quadratic irrationals.

Greeks obtained these numbers as an answer to the following geometric question "Given a line AB of unit length, obtain the point C within AB, such that (AB/AC) = (AC/BC)". If we denote the length AC by ϕ, then we get to satisfy the proposed requirement

$$\frac{1}{\phi} = \frac{\phi}{1 - \phi} \text{ or, } \phi^2 + \phi - 1 = 0 \tag{3.27}$$

Solving (3.27) and choosing the positive root, we get $\phi = \frac{\sqrt{5} - 1}{2}$. [Because the answer was known, we used the symbol ϕ for AC in advance]. Finally, we can write

$$\psi = \frac{1}{\phi} = 1 + \phi = \frac{\phi}{1 - \phi} \qquad (3.28)$$

Similar to the golden section of a straight line of unit length, one defines golden angle with respect to a circle of unit perimeter. The total angle at the centre subtended by the perimeter of a circle is 2π, the golden angle (θ) is defined such that

$$\frac{2\pi}{2\pi - \theta} = \frac{2\pi - \theta}{\theta} \qquad (3.29)$$

Solving (3.29) and choosing the value less than 2π, we get

$$\theta = \frac{3 - \sqrt{5}}{2}(2\pi) = (1 - \phi)2\pi = \phi^2(2\pi) \qquad (3.30)$$

The reader is advised to check that the infinite continued fraction (0; 2, 1, 1, 1, ...) expresses the value of ϕ^2.

The definition of *golden* entity was extended to many other geometrical shapes like, rectangle, parallelepiped, ellipse, spiral and so on. In these objects, ratios of the defining length parameters was taken as ϕ. Several books have been written on this *divine proportion*. Here we discuss only the golden rectangle and the golden spiral and mention some of their interesting mathematical properties. Greek artists and architects considered the golden rectangle as aesthetically the most pleasing of all rectangles. In some sense, they were obsessed with the golden proportion. Greek geometers located golden rectangles within Platonic solids, which are regular solids bounded by identical regular polygonal surfaces. It has been proved that there exist only 5 such solids (See Appendix E).

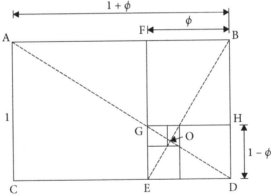

Fig. 3.3: Golden rectangle

The rectangle ABCD, shown in Fig. 3.3, is a golden rectangle where the ratio of the sides (AC/AB) is ϕ; consequently the ratio (AB/AC) is $(1/\phi) = \psi = 1 + \phi$, as indicated. From this golden rectangle if the unit square AFEC is removed, the remaining rectangle FBDE is also a golden rectangle. This process of removing the square and generation of smaller and smaller golden rectangles continues forever, until the limiting point O is reached. The point O is at the intersection of diagonals AD and BE of the first two rectangles. These two lines AD and BE remain, respectively, the diagonals of all golden rectangles of same orientation. The ratio (BE/AD) is also ϕ.

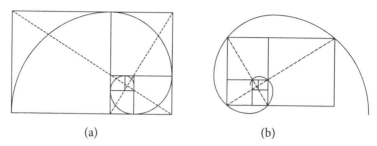

(a)   (b)

Fig. 3.4: Golden spiral (a) counterclockwise (b) clockwise

Starting from a golden rectangle nesting of infinite number of rectangles can start from the left (as shown in Fig. 3.3) or from the right. Figs 3.4(a) and (b) show both these types of nesting. The spiralling curves drawn through the corners of the squares shown in these two figures are called *golden spirals*. The first one is spreading out around the origin O (in Fig. 3.3) in counter-clockwise (CCW) direction and the second one is spreading out in the clockwise direction. Let us focus on the CCW one. The shape of the curve does not change only its size increases as it spreads outward. The portion of the curve within any square if magnified by a factor ψ and rotated through 90° in the CCW direction, one gets the entire curve within the adjacent square in the CCW direction. This curve is a special case of a class of generalized curves called *equiangular* or *logarithmic spirals*. The equiangular spiral has very interesting properties, which were extensively studied by Jacob Bernoulli.

Using polar coordinates, with the origin at O, the equation of these curves can be written as $r = r_0 e^{\gamma\theta}$. We can set $r_0 = 1$ (i.e. the scale of the figure) by defining $r = 1$ at $\theta = 0$ when the equation is simplified as $r = e^{\gamma\theta}$. Fig. 3.5 shows an equiangular spiral. At any point on this curve, the angle between the radius vector from the origin and the tangent at that point remains equal (α), hence the name equiangular. It can be shown $\alpha = \cot^{-1}\gamma$, with $\gamma > 0$ for CCW spiral. Note that, as is well known, with $\alpha = 90°$, $\gamma = 0$; this makes r constant, i.e. the spiral becomes a circle. It can be also shown (see Fig. 3.5), from any point P the length of the curve from P to the origin is given by PT, where T is the point of intersection of the tangent at P and the line through O perpendicular to OP. It is easy to see that $PT = (r/\cos\alpha)$. The reader is asked to prove that for the golden spiral

$$\gamma = \frac{2}{\pi} \ln \psi = \frac{2}{\pi} \ln \frac{\sqrt{5}+1}{2} \tag{3.31}$$

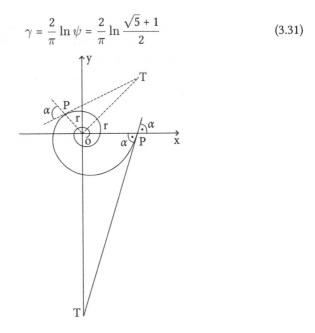

Fig. 3.5: Equiangular or logarithmic spiral

Now we discuss curious connections between the golden ratios and the Fibonacci (Hemachandra) sequence discussed in Section 2.9. Towards this end, first we use (2.31) to calculate the limiting value ϕ (we use this symbol, because the answer is known to the author) of the two consecutive numbers in this sequence as $n \to \infty$. We write

$$\frac{F_n}{F_{n-1}} = 1 + \frac{F_{n-2}}{F_{n-1}} \text{ or, } \frac{1}{\phi} = 1 + \phi \text{ or, } \phi^2 + \phi - 1 = 0 \tag{3.32}$$

which is same as (3.27), thus the use of the symbol ϕ is justified. Thus, the Fibonacci sequence is closely related to the golden sections. In fact, the nth Fibonacci number F_n can be written in terms of ψ as given below:

$$F_n = \frac{\psi^n - (-\psi)^n}{\sqrt{5}} \tag{3.33}$$

It may be worthwhile at this stage to mention some interesting facts regarding frequent appearances of Fibonacci numbers, and different golden entities in the self made tapestry of nature consisting of the plant and animal worlds. Such mathematical descriptions, and sometimes explanations of the proportions and shapes found in the living world are becoming quite important in botany and biology. Without going deep into this vast field we just give some common examples which most of us have come across but probably did not pay attention.

We start with the number of ancestors encountered in different generations of a male (M), or worker honeybee. It may be recalled that these workers or male honeybees are produced directly from the larva of a female (F) honeybee. The larva is not fertilized; hence a male honeybee has only a mother and no father. But a female honeybee has both a mother and a father. In other words, a male honeybee has only one female (F) ancestor in the immediate previous generation. In Table 3.1, we start with a male honeybee and follow its ancestors in previous generations. We go downward one step for the immediately previous generation and indicate a father or Male by M and a mother or Female by F. It is clearly seen that the total number of honeybees in every generation follow the Fibonacci numbers. The same statement holds for the females and males if we start counting, respectively from one and two generations back.

Table 3.1

		M		
		F		
	M		F	
	F	M	F	
M	F	F	M	F
F	MF	MF	F	MF

....

....

It has also been found that the number of petals in more than 90 percent flowers is one of the Fibonacci numbers, like, 1, 2, 3, 5, 8, 13, 21, 55, 89, ... In daisies we find 55 petals and in giant sunflowers 89 petals are found. It has been reported that so far all the flower petals have been found to belong to one of the three sequences, viz. the Fibonacci sequence or a similar sequence with starting numbers 4, 7, i.e. 4, 7, 11, 18, 29, ... or the double of Fibonacci numbers like 2, 4, 6, 10, 16, 26, 42, ... It implies you cannot find a flower with number of petals which does not belong to any of these three sequences, e.g. 12 or 56, etc.

We know that the skin of a pineapple exhibits helical *whorls*, called *parastichies*. As the pineapple grows in diameter and height, more and more whorls appear and arrange themselves along two oppositely directed helical spirals as shown in Fig. 3.6. The numbers of such spirals are two consecutive Fibonacci numbers like (5, 8) or (8, 13) depending on the size of the pineapple. Let these approximately hexagon shaped whorls be numbered chronologically starting from 1, as they are born. Now observe the numbers on the hexagonal cells adjacent to a given one. The difference with that on the given one is seen to be 5, 8 or 13 depending on which two opposite and parallel sides of the hexagonal cell are used to define the adjacent cell. Surprisingly, again 5, 8 and 13 are three consecutive Fibonacci numbers.

Fig. 3.6: Whorls on the skin of a pineapple

Now again without going into the details, let us consider the appearance of golden angles in phyllotaxis or leafing pattern and deposition of florets in a sunflower. As a tree grows, leaves come out of the growing stem and the angle of deposition of the consecutive two leaves is such that the sunlight is not obstructed to reach the earlier (lower) leaves. Table 3.2 shows the fraction of a complete turn between the angles of deposition between two consecutive leaves of various trees. It is seen that for the high trees this ratio is (1/2), and for very short trees, like that of potato, it tends to the golden angle given by (3.27). The other values listed in Table 3.2 are seen to be the rational approximation (convergent) of the continued fraction expression of the golden angle given immediately after (3.30). These are given by the ratios of two alternate (not consecutive, i.e, with a gap of one) numbers in the Fibonacci sequence.

Table 3.2

Type of tree	Leaf deposition angle as a fraction of a complete turn
Elm, Linden	1/2
Beech, hazel	1/3
Oak, Apple	2/5
Poplar, Rose	3/8
Willow, Almond	5/13

Fig. 3.7: Double spiral of florets on a sunflower

Fibonacci numbers and the golden angle also appear in the sunflower, shown in Fig. 3.7. Two oppositely directed spirals are clearly seen in the pattern of florets. Florets are deposited on a spiral (not shown in the figure) as the sunflower grows in size. It has been shown that the deposition of consecutive florets has to be at a separation of the golden angle for the ultimate generation of this double spiral pattern. It is obvious that if the deposition occurs at a rational fraction of a complete turn, then radial spikes of florets are generated. It has been shown that if the deposition occurs at an angle different from the golden angle, then the double spiral pattern is limited to a small central region, followed by helical spikes in the outer region. The helix turns counterclockwise if the angle of deposition is less than the golden angle and turns clockwise otherwise. Needless to say, this compact double spiral pattern is essential for the flower to withstand the force exerted by the wind.

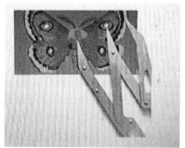

Fig. 3.8: Golden mean gauge

Not only the Greek artists and architects gave a lot of importance to the aesthetics of golden quantities. Even today there is a controversial opinion that the best ratio of the height up to the navel to the total height of a beautiful human figure is the golden section, ϕ. In fact, an orthodontist and a cosmetic surgeon in London, named Eddy Levin[1], believes that the most beautiful ratio, of the widths

---

[1] Aesthetic dentistry today, May 2011, vol. 5, no. 3, p. 22.

and spacing between different teeth, of some facial dimensions, is the golden section. In fact, for quick measurement of the golden ratio, he has devised and patented a three legged caliper, where the distances between the three legs always maintain the golden ratio. Such a caliper is shown in Fig. 3.8 to demonstrate that a butterfly is beautiful because of the existence of golden ratio between different dimensions on its body!

We conclude this section by briefly mentioning the logarithmic spiral and logarithmic helico spirals (in three dimensions) that are seen in the shapes of various objects which grow with time. These include spiral nebula, conch shells, chambered nautilus, and horns of a mountain sheep and so on. We have already mentioned that golden spiral is a logarithmic spiral. And the characteristic of such spirals is that the curve just grows in size without altering the shape. Nature probably prefers this optimization of effort. Very simple mathematical models of growth of the objects mentioned above have been proposed in mathematical biology.

3.9.2 Oldest universal mathematical constant π

Greek geometers proved that the circumference of a circle is proportional to the diameter and the area of a circle is proportional to the square of its radius. They also knew that these two constants of proportionality have the same numerical value. Great mathematician Euler, through his enormous works, made the symbol π an universal one for this oldest mathematical constant. Different approximate values of this constant were reported in different ages, in different civilizations. Plato used $\pi \approx \sqrt{2} + \sqrt{3} \approx 3.14626\ldots$ Archimedes determined the approximate value of π by calculating the perimeters of an inscribed and a circumscribing 96-sided polygon in and around a given circle. He obtained $3\frac{10}{71} < \pi < 3$. The great Indian mathematician Aryabhatta used the fraction (62382/20000) as the value of π. In school mathematics, the most popular values are 22/7 as a rational number and 3.14 in decimal notation. There are other not so uncommon values also, such as 355/113, as mentioned in Section 3.5. Ramanujan gave a geometric construction for 355/113 starting from a circle of unit radius.

There is a hilarious story regarding the standard value of π. The house of representatives in the Indiana state of USA once made an act *introducing a new mathematical truth* proposing two values, viz. 4 and 3.2 for π. Fortunately, the adoption of this act was postponed indefinitely. In the second half of the nineteenth century, it was proved that π is a transcendental number; this proof is a great milestone in the history of mathematics. Since then, using digital computers a trillion digits after the decimal point have been obtained. These calculations have been carried out using infinite series connected to the value of π. In what follows, we just mention a few of these, which are important in the history of mathematics.

Madhava (1350–1425) — Gregory (1671) — Leibniz (1673) series

$$\frac{\pi}{4} = \frac{1}{1} - \frac{1}{3} + \frac{1}{5} - \frac{1}{7} + \dots \tag{3.34}$$

This apparently simple series is not useful for determining the value of π, as 500 terms are necessary to correctly obtain the first three digits after the decimal point. In Europe, the first series was given by Vieta (1579) in the form

$$\frac{2}{\pi} = \sqrt{\frac{1}{2}} \sqrt{\frac{1}{2} + \frac{1}{2}\sqrt{\frac{1}{2}}} \sqrt{\frac{1}{2} + \frac{1}{2}\sqrt{\frac{1}{2} + \frac{1}{2}\sqrt{\frac{1}{2}}}} \tag{3.35}$$

In 1737 Euler was recognized as the best mathematician in Europe when he proved

$$\frac{1}{1^2} + \frac{1}{2^2} + \frac{1}{3^2} + \dots = \frac{\pi^2}{6} \tag{3.36}$$

The series (3.36) is useless for approximating the value of π, as 100 million terms are needed to correctly obtain the first seven digits after the decimal point. However, obtaining this sum in closed form was a grand feat as it stumped all earlier great mathematicians. Moreover, this result has also given rise to a huge development of mathematics in other areas.

The great Indian genius Ramanujan gave the following surprising result

$$\pi^4 = 97 + \cfrac{1}{2 + \cfrac{1}{2 + \cfrac{1}{3 + \cfrac{1}{1 + \cfrac{1}{16539 + \dots}}}}} \tag{3.37}$$

The above continued fraction results in

$$\pi^4 \approx 97\frac{9}{22} \quad \text{or, } \pi \approx 3.14159265258\dots \tag{3.38}$$

which correctly gives the first eight digits after the decimal point. Starting with a circle of unit radius, Ramanujan also provided a geometric construction for obtaining the above value. In 1910, Ramanujan gave the following series for the evaluation of π.

$$\frac{1}{\pi} = \frac{2\sqrt{2}}{9801} \sum_{k=0}^{\infty} \frac{(4k)!(1103 + 26390k)}{(k!)^4 396^{4k}} \tag{3.39}$$

The first term of (3.39) gives correctly the first six digits after the decimal point; each additional term gives 9 more digits accurately. In 1994, two Russian mathematicians modified (3.39) as

$$\frac{1}{\pi} = 12 \sum_{k=0}^{\infty} \frac{(-1)^k (6k)!(545,140,134k + 13,591,409)}{(3k)!(k!)^3 640,320^{3k+\frac{3}{2}}} \tag{3.40}$$

The first term of the series (3.40) correctly gives the first 13 digits after the decimal point and each additional term gives 14 more digits accurately. In 2013, a trillion digits after the decimal point were computed using (3.40).

3.9.3 Base of natural logarithm e

After π, the most widely used universal number is e. This symbol was also popularized by Euler through his enormous amount of mathematical work and now it is used universally. This number, which we meet in calculus, is defined by the following limit:

$$e = \lim_{n \to \infty} \left(1 + \frac{1}{n}\right)^n = 2.71828182845\ldots \tag{3.41}$$

The expression, whose limit is considered above, we encounter much earlier in arithmetic while considering compound interest. Let us consider a principal amount of ₹ 1, loaned at an annual 100 % rate of compound interest. Let us answer the question what is the maximum amount of money that can be got back after 1 year? Obviously the answer will depend on how many times (n) the interest will be calculated at an interval of $(1/n)$ year. Table 3.3 shows the money received after 1 year for various values of n.

Table 3

Value of n	Amount of interest	Total money received
1	1	$1 + 1$
2	$\frac{1}{2}\left(1 + \frac{1}{2}\right)$	$\left(1 + \frac{1}{2}\right)^2$
3	$\frac{1}{3}\left(1 + \frac{1}{3}\right)^2$	$\left(1 + \frac{1}{3}\right)^3$
$\vdots$	$\vdots$	$\vdots$
n	$\frac{1}{n}\left(1 + \frac{1}{n}\right)^{n-1}$	$\left(1 + \frac{1}{n}\right)^n$
$\vdots$	$\vdots$	$\vdots$
∞	$e - 1$	$e = 2.71828182845\ldots$

So the maximum amount that can be received, even if the interest is calculated infinite times, i.e. continuously at every instant, cannot exceed ₹ 2.718... approximately ₹ 2.72.

With the introduction of the exponential function defined as e^x, with x as the independent variable, the symbol e became ubiquitous in various branches of mathematics. The exponential function is the only function which is its own derivative. In the form of infinite series, this function is written as

$$e^x = 1 + \frac{x}{1!} + \frac{x^2}{2!} + \frac{x^3}{3!} + \ldots, \quad -\infty < x < \infty \tag{3.42}$$

Substituting $x = 1$ in (3.42), we get another definition of e as

$$e = 1 + \frac{1}{1!} + \frac{1}{2!} + \frac{1}{3!} + \ldots \tag{3.43}$$

Isaac Newton (1642–1727, born in 1643 according to modern Gregorian calendar) obtained the very fast convergent series (3.41) for the number e by binomial expansion of $(1 + \frac{1}{n})^n$ with $n \to \infty$.

You are already familiar with the logarithm of base 10, $\log_{10} x$, where powers of 10 were considered. Following this, natural logarithm $\ln x$ is defined with e as the base, i.e.

$$e^{\ln x} = x \tag{3.44}$$

So the number e is the base of the natural logarithm.

Before proceeding any further, let us answer the question raised regarding the solution of (3.13). The process of successive exponentiation (from right to left) of a real number x was carried on infinite times or forever. If the process of successive exponentiation, called *tetration* (inverse operation of taking root), is carried over n times then it is written as $^n x$, the limit of $^n x$ exists as $n \to \infty$, only if $e^{-e} \le x \le e^{1/e}$ and the values of $^n x$ at the extremes of this range are, respectively $(1/e)$ and e. We note that $e^{-e} \approx 0.065988035 \ldots$, $e^{1/e} \approx 1.444667861 \ldots$, $(1/e) \approx 0.367879441 \ldots$ and $e \approx 2.718281828 \ldots$

Now returning to (3.13), we see that the righthand side cannot be 4, which is $> e$, and $\sqrt{2} \approx 1.414 \ldots$, which is within the range $e^{-e} \le x \le e^{1/e}$, so the lefthand side exists. In fact the righthand side is 2 which lies within the range $(1/e)$ and e.

3.9.4 Euler's constant γ

Another very important real number, believed to be transcendental, but not proved to be so, is called *Euler's constant*, with the universal symbol γ. This number, used in various areas of mathematics, has been calculated up to more than 100 million decimal places. No periodic appearance of the digits has been observed. This number was first discussed by Euler in the context of the sum of a harmonic series consisting of the reciprocals of all the consecutive natural numbers. In what follows, we first briefly discuss the harmonic series.

Let the harmonic series up to n terms be defined as

$$H_n = \frac{1}{1} + \frac{1}{2} + \frac{1}{3} + \cdots + \frac{1}{n} \tag{3.45}$$

Unlike the sum of the first n natural numbers, no closed form expression for H_n could be obtained. French polymath Nicolas Oresme (1323–82) provided a simple proof that the limit of the sum H_n as $n \to \infty$ diverges. Though it may appear counter intuitive as we add smaller and smaller term, let us just write

$$
\begin{aligned}
H_\infty &= \frac{1}{1} + \frac{1}{2} + \frac{1}{3} + \frac{1}{4} + \frac{1}{5} + \frac{1}{6} + \frac{1}{7} + \frac{1}{8} + \dots \\
&> 1 + \frac{1}{2} + \left(\frac{1}{4} + \frac{1}{4} \right) + \left(\frac{1}{8} + \frac{1}{8} + \frac{1}{8} + \frac{1}{8} \right) + \dots \\
&> 1 + \frac{1}{2} + \frac{1}{2} + \frac{1}{2} + \dots
\end{aligned}
\tag{3.46}
$$

So H_∞ obviously diverges as we go on adding $(1/2)$ indefinitely forever. The harmonic series diverges very slowly. Euler tried to answer the question *how fast this infinite series diverges?* For this purpose he looked at the difference between H_n and $\ln n$ with increasing n. Table 3.4 shows a few such values.

Table 3.4

n	H_n	$\ln n$	$H_n - \ln n$
100	5.187 ...	4.606 ...	0.581 ...
1,000	7.486 ...	6.909 ...	0.575 ...
1,000,000	14.392 ...	13.818 ...	0.574 ...

Euler proved that the difference, $\gamma_n = H_n - \ln n$, reaches a limit as $n \to \infty$. This limit, $\gamma = \lim_{n \to \infty} \gamma_n$ is called the *Euler's constant.* Euler also calculated the value of γ as $0.5772156649015325...$, which is correct up to 15 decimal places. In some places this constant is mentioned as Euler-Mascheroni constant. Mascheroni also computed the numerical value of γ up to 32 decimal places; later on it was pointed out that the 20th digit was wrong. Euler used the even Bernoulli numbers (discussed in Section 3.8.1) to write the following formula

$$
\gamma \approx \gamma_n - \frac{1}{2n} + \sum_{m=1}^{\infty} \frac{B_m}{2m} \frac{1}{n^{2m}}
\tag{3.47}
$$

We may recall that Bernoulli numbers were defined in Section 3.8.1 in an altogether different context. The appearance of the same numbers in (3.47) is another example of curious connections between different types of numbers.

Chapter 4

Problems

In Problem Nos 1–9 below, crosses (X) represent missing digits which may or may not be repeated. Each of the letters A, B, C, ..., of the English alphabet represents a different digit within the same problem.

1. Determine the digits represented by A and B if the number A 4273 B is divisible by 72.

2. Determine the missing digits in 20! = 24329020081XX640000.

3. In the division process shown below except one 8 all other digits are missing and there is no remainder. Find all the missing digits.

$$
\begin{array}{r}
\text{X X X) X X X X X X X X X (X X 8 X X} \\
\text{X X X} \\
\hline
\text{X X X X} \\
\text{X X X} \\
\hline
\text{X X X X} \\
\text{X X X X} \\
\hline
\end{array}
$$

In Problems 4–9, determine the digits represented by various letters to make the equations correct.

4. A B C + B C A + C A B = A B B C.

5. S B N D + M O R E = M O N E Y.

6. F O R T Y + T E N + T E N = S I X T Y.

7. A B C D × 4 = D C B A.

8. A B C D E F × 5 = F A B C D E.

9. A B C D E × A = E E E E E E.

10. Find the number of zeroes (0) at the end of the number 98! + 99! + 100!.

11. Given that the number N! has exactly 1993 zeroes (0) at the end. Determine all possible values of N.

12. (a) Obtain the highest possible value of n so that 5^n is a factor of 100!.

(b) Obtain the highest possible value of n so that 6^n is a factor of 120!.

13. Find the last two digits of $11^{10} - 1$.

14. How many perfect square factors exist for 12!?

15. Let $N^2 = 25^{64} \times 64^{25}$. What is the sum of all the digits of the positive integer N?

16. How many numbers are divisors of both 7560 and 8400?

17. What is the smallest number having exactly 24 divisors?

18. How many divisors of 2160 are even?

19. How many divisors of 2160 are divisible by 3?

20. (a) A positive integer N has two prime factors and total nine divisors. How many total divisors does the number N^2 have?

    (b) A number that is a perfect cube has 28 divisors. The cube root of that number has N divisors. What is the sum of all possible values of N?

    (c) What is the smallest number which has exactly 100 divisors?

21. (a) Each of the three natural numbers N_1, N_2 and N_3 has exactly three divisors. How many total divisors does the number $N_1^2 N_2^3 N_3^4$ have?

    (b) Two concentric circles have diameters given by two different integers. The area of the annular region between these two circles is 36π. What is the sum of all the possible values of the diameters of the smaller circle?

    (c) The sum of the reciprocals of two different positive integers is 1/12. What is the sum of all possible values of the smaller number?

    (d) If a, b and c are positive whole numbers such that $ab = 4a + 2b + 34$ and $bc = 5b + 3c + 34$, then what is the value of $a + b + c$?

22. For which positive integers n, the number $(n^3 - n)$ is divisible by $(n - 3)$?

23. Find S^2, when $S = 333 \ldots 333$, where the digit 3 has been written 333 times.

24. (a) Determine the sum of the digits in $S = 1+11+111+1111+\cdots+111\ldots111$, where the last term contains 111 ones.

    (b) A palindrome number remains the same when read forward or backward. What is the sum of all three-digit palindrome numbers?

    (c) How many three digit palindrome numbers are not divisible by 11?

25. Consider all six digit numbers of the form $abccba$, where it is known that the digit b is odd and the digits a, b and c may or may not be different. How many of these numbers are divisible by 7?

26. (a) Prove that the number $5555^{2222} + 2222^{5555}$ is divisible by 7.

    (b) What is the remainder when 65^{2021} is divided by 63.

27. Let a number $N = 6 + 66 + 666 + \cdots + 666, \ldots 666$, where in the last term 6 is repeated 100 times. If we write the number N explicitly how many times can the digit 7 be seen in it?

28. Prove that for all positive integers n, the number $(n^5 - n)$ is divisible by 5.

29. (a) Prove that the product of four consecutive natural numbers cannot be a perfect square.

    (b) What is the sum of the digits of $\sqrt{1 + 2019 \times 2020 \times 2021 \times 2022}$?

30. Prove that all perfect square numbers $\gg 1$) are either divisible by 4 or leave a remainder 1 when divided by 8.

31. Let m and n be two natural numbers such that the three numbers $(2m - n)$, $(m - 2n)$ and $(m + n)$ are all different perfect squares. Determine the minimum possible value of n.

32. Prove that for all primes $p > 3$, the number $(p^2 - 1)$ is divisible by 24.

33. Prove that for two natural numbers m and n, satisfying the relation $m = n^2 - n$, the number $(m^2 - 2m)$ is divisible by 24.

34. The product of the digits of a natural number n is given by $n^2 - 15n - 27$. Determine all the possible values of n.

35. Prove that the numbers written below using only the digit 1 cannot be perfect squares: 11, 111, 1111, 11111, ...

36. Prove that for all odd natural numbers n, the number $(n^4 + 4n^2 + 11)$ is divisible by 16.

37. Three integers satisfy the following two equations:

$$p + q - r = 1$$
$$p^2 + q^2 - r^2 = -1$$

Determine the sum of all possible values of the expression $(p^2 + q^2 + r^2)$.

38. Prove that a natural number n can be expressed as a sum of two squares if and only if the number $2n$ can be expressed as a sum of two squares.

39. Let n be an integer, negative, zero or positive. Determine the values of n for which the number $(n^2 + 19n + 47)$ is a perfect square.

40. Consider all the ten-digit numbers consisting of four 4's, three 3's, two 2's and one 1. How many of these are perfect squares?

41. Six natural numbers p, q, r, s, t and u are in strictly increasing order. The fifteen pair wise sum of these numbers are 25, 30, 38, 41, 49, 52, 54, 63, 68, 76, 79, 90, 95, 103 and 117. Which one is representing the sum $(r + s)$?

42. What is the least number which when it divided by 5, 7 and 8 leaves a remainder 1?

43. What is the least number which when divided by 5 leaves a remainder 2, when divided by 8 leaves a remainder 5 and when divided by 11 leaves a remainder 8?

44. (a) What is the largest number which when it divides 364, 414 and 539 leaves the same remainder.

    (b) What is the greatest three-digit number which when divided by 8 leaves a remainder 2, when divided by 6 leaves a remainder 4 and when divided by 7 leaves a remainder 3?

    (c) What is the least number which when divided by 23 leaves a remainder 2, when divided by 232 leaves a remainder 23 and when divided by 2323 leaves a remainder 232?

45. How many odd four-digit numbers with distinct digits can be written?

46. How many four-digit multiples of 5 can be written having distinct digits with 5 as the greatest digit?

47. A company decided to issue identity cards to their employees with six-digit numbers which must have at least two different digits. How many such cards can be issued and how to ensure that the desired requirement is fulfilled?

48. In writing all the numbers from 1 to 100,000 how many total digits are used?

49. How many five-digit multiples of 3 are there where the digit 6 is seen to appear at least once?

50. A six-digit display meter runs to show all the numbers 000,000 to 999,999. How many numbers showed the digit 1 at least once? How many 1's could be seen during this display? Note in 000101 we can see two 1's.

51. Seven different positive integers add up to 100. Show that there must be three numbers which add up to at least 50.

52. Consider a set of n positive integers $(a_1, a_2, a_3, \ldots, a_k)$, not necessarily distinct. Now consider all possible subsets formed using these numbers. Prove that there must be at least one subset the sum of whose element(s) must be divisible by n.

53. You ask a friend to write a polynomial, out of your sight, in a single variable x as

$$P(x) = \sum_{i=0}^{n} a_i x^i$$

where all the coefficients a_i's are non-negative integers. You ask him the value of the polynomial for two values of x of your choice. From this information you have to determine the entire polynomial, i.e. the value of the order n and all the coefficients. Describe a scheme to to achieve this and demonstrate for the example $P(x) = 1 + 12x + 2x^3$.

54. Consider a polynomial

$$P(x) = \sum_{i=0}^{n} ax^i$$

where all the coefficients a_i's are non-negative integers. Given $P(1) = 4$ and $P(5) = 136$, determine the value of $P(3)$.

55. If the number 2^{29} is written explicitly then nine different digits all appear once. Which digit does not appear?

56. Determine the value of the following product

$$\frac{7}{9} \times \frac{26}{28} \times \frac{63}{65} \times \cdots = \prod_{m=2}^{\infty} \frac{m^3 - 1}{m^3 + 1}$$

57. Write the number N as a single number, where

$$N = \frac{(10^4 + 324)(22^4 + 324)(34^4 + 324)(46^4 + 324)(58^4 + 324)}{(4^4 + 324)(16^4 + 324)(28^4 + 324)(40^4 + 324)(52^4 + 324)}$$

58. Make 4 pieces out of a 40 kg stone so that these can be used in a standard two pan balance to weigh all integer kilograms.

59. (a) Determine the sum of the following infinite series

$$S = \frac{1}{1} + \frac{1}{3} + \frac{1}{6} + \frac{1}{10} + \frac{1}{15} + \cdots + \frac{2}{n(n+1)} + \cdots$$

(b) Express the value of

$$\frac{3}{2^2} + \frac{5}{6^2} + \frac{7}{12^2} + \frac{11}{30^2} + \frac{13}{42^2}$$

as a fraction in the lowest form.

60. First guess an answer for the following series by considering the first few terms and then prove that your guess is correct: $S_n = 1!1 + 2!2 + 3!3 + \cdots + n!n$.

61. First guess an answer for the following series by considering the first few terms and then prove that your guess is correct:

$$S_n = \frac{1}{2!} + \frac{2}{3!} + \frac{3}{4!} + \cdots + \frac{n}{(n+1)!}$$

62. Evaluate the following infinite series

$$S = \sum_{i=1}^{\infty} \frac{1}{i(k+1)}$$

63. For all natural numbers n, prove the following relation

$$\frac{1}{1.2} + \frac{1}{3.4} + \frac{1}{5.6} + \cdots + \frac{1}{(2n-1)2n} = \frac{1}{n+1} + \frac{1}{n+2} + \frac{1}{n+3} + \cdots + \frac{1}{2n}$$

64. (a) Prove the following relation

$$3 = \sqrt{1 + 2\sqrt{1 + 3\sqrt{1 + 4\sqrt{1 + \ldots}}}}$$

(b) Find the sum of the digits of the following product:

$$\underbrace{999,999,\ldots,999,999}_{\text{9's appear 93 times}} \times \underbrace{444,444,\ldots,444,444}_{\text{4's appear 93 times}}$$

65. A ten digit number $N = abcdefghij$ uses all the ten digits in such a way that a is divisible by 1 (trivially true), the number ab is divisible by 2, the number abc is divisible by 3, the number $abcd$ is divisible by 4 and so on and finally the number $abcdefghi$ by 9 and the entire number N by 10. Find the number N.

66. Determine two integers x and y satisfying the equation $x^2 + 615 = 2^y$.

67. The 10 real numbers a_i's with $i = 1, 2, 3, \ldots, 10$ are such that for all values of i, $a_i < a_{i+1}$; Prove that

$$\sum_{i=1}^{6} a_i \Big/ 6 < \sum_{i=1}^{10} a_i \Big/ 10$$

68. (a) Consider seven real numbers y_i, with $i = 1, 2, 3, \ldots, 7$, in the range $-\infty$ to $+\infty$. Prove that some two of these, say y_j and y_k must satisfy the follow condition:

$$0 \le \frac{y_j - y_k}{1 + y_j y_k} \le \frac{1}{\sqrt{3}}$$

(b) Consider all the natural numbers 1 to 100. Pick up a set of 10 numbers from these 100 numbers. Prove that there must be two disjointed subsets of this set of 10 numbers which have the same sum.

69. Find the remainder when $(x^{2013} + 1)$ is divided by $(1 + x + x^2)$.

70. Find all the real values of x satisfying the following equation where n is a natural number:

$$\sum_{k=1}^{n} \frac{kx^k}{1 + x^{2k}} = \frac{n(n+1)}{4}$$

71. Find all the real values x satisfying the following equation:

$$(x^2 - 7x + 11)^{(x^2 - 13x + 42)} = 1$$

72. Express 271 as a sum of positive real numbers so that the product of these is maximized.

73. On June 1 all the students of a class were asked to stand in 1 row. On June 2 the students were asked to stand in one column, on June 3 in 5 rows, on June 4 in 3 rows, on June 5 in 6 rows and on each successive days they were asked to stand in a different number of rows than on earlier days. The process continued until June 12. Assume no holidays and no student is ever absent. On June 13 the teacher found that no new formation is possible, he has to use one of the earlier formations. What is the minimum number of students in this class?

74. 200 students, whose heights are all different, are arranged in 10 rows and 20 columns. First the shortest person from each column is chosen and the tallest of them is marked as A. They are asked to return to their original positions. Then the tallest of each row is chosen and the shortest of them is identified as B. If A and B are different who is taller among these two? Explain your answer.

75. On December 31, 2016 a grandmother received phone calls from her three granddaughters. The eldest one calls her every three days, the youngest on every 5 days and the middle on every 4 days. On how many days during the year 2017 did she not receive any call from her granddaughters?

76. In the barrack of a jail, 100 cells in one line were sequentially numbered from 1 to 100. One night a whimsical guard visited all the cells consecutively, starting from number 1, opened the locks and returned to the starting point. He then visited every alternate cell starting from 2 and changed the state of the locks; i.e. opened the closed ones and locked the open ones. Then he returned after reaching the end of the line. Now he started from cell 3 and visited every third cell and changed the state of the locks. This process continued until he had to start from the last cell. At the end of this process which cells remained open?

77. 10 girls are standing in a circle at equal distances. Each one tells a number to her two immediate neighbours. Thus, every one gives one number and receives two numbers from her two neighbours. Surprisingly when they announced the average of the two numbers they received, they said 1, 2, 3, ..., 10 in a sequential manner. What number was given by the girl who announced 6?

78. Two friends A and B start running simultaneously from two diametrically opposite points on a circular track. They run in opposite directions at different but uniform speeds. When they first meet A has run 300 m. They continue running and when they meet again, B has run 400 m between their first and second meetings. What is the length of the track?

79. Three friends A, B and C start walking simultaneously from the same location in the same direction at different but uniform speeds. When A walks 900 m, B walks 800 m and C walks 700 m. A takes 5 minutes to complete 1 round of this 1 km long track. After how many minutes are all three at the same location for the first time after starting?

80. 1000 people stand along a circle at equal distance. Call any of the places arbitrarily as 1. Now moving in the clockwise direction starting with the second person every alternate person goes away until only the last person remains. What was the original place number for this person?

81. In problem number 80 what will be the answer if starting with the third person, every third person (i.e. 3, 6, 9, ...) goes away?

82. How many different ten-digit numbers can be written with four 4's, three 3's, two 2's and one 1?

83. You have 4 Bengali, 3 English and 2 Hindi books. All books are different. How many different ways can you put these books on your book shelf if you want to keep all the Bengali books together?

84. How many different arrangements with the letters of the word BABOON are possible if 2 O's are to be always adjacent and 2 B's must be separated by at least one other letter?

85. A book starts with page number 1. In total 3189 digits have been used to print all the page numbers. How many pages are there in this book?

86. A book is published in 3 volumes. The page number starting with 1 on the first page of volume 1 runs continuously through all the volumes. The second volume has 50 more pages than the first. The third volume contains one and a half times as many pages as the second volume. The sum of the starting page numbers of the three volumes is 1079. What is the largest prime factor of the last page number of the book?

87. One leaf of a page from a *normal book* has been torn. The page numbers on the remaining pages add up to 15,000. What are the page numbers on two sides of the torn leaf?

88. (a) A bag contains 20 black and 16 white balls. Sufficient numbers of similar balls are also available outside. Two balls are taken out of the bag at random without replacement; but if these are of the same colour then a black ball is put inside. If the two balls taken out are of different colours, then a white ball is put inside the bag. Thus, at every step the number of balls inside the bag reduces by one. If the process continues until only one ball is left what is the colour of that ball. Explain your answer.

(b) A staircase consists of 12 stairs. You can either go up 1 step at a time or skip one stair (i.e. jump over 1 stair). In how many different ways can you reach the top of the staircase?

89. First you write all the numbers 1, 2, 3, …, 2018 on a piece of paper. Now erase any two arbitrarily chosen numbers say a and b. Then write a new number given by $|a - b|$. Continue this process until only one number is left. Can this number be zero? Explain your answer.

90. Five people have to cross a narrow bridge on a dark night. For moving on the bridge it is essential to use the single torch they have with them. At a time a maximum of two people can walk over the bridge. These people walk at different speeds and take, respectively 1, 3, 6, 8 and 13 minutes to cross the bridge. Obviously when 2 people go together with the single torch, the speed is dictated by the slower of the two persons. The battery of the torch has only 30 minutes of life left. Devise a scheme for the group to cross the bridge.

91. A person gets down at his station at 4 pm every day. His wife brings the car exactly at that time to pick him up and drives him back home. One day he reached the station at 3 pm and started walking towards home. He met his wife on the way and got into the car and returned home to find that he has reached 10 minutes earlier than usual. How many minutes did he walk? Assume no time is wasted anywhere and all speeds involved are uniform.

92. A steamer goes downstream from A to B in 5 hours and returns upstream from B to A in 7 hours. Assume constant speeds for the steamer engine and the current. How long will a wooden plank take to reach from A to B flowing with the river current? This problem has to be solved without using any paper and pen.

93. On a still river two boats start simultaneously from opposite banks and while crossing the river meet each other at a distance of 800 m from one bank (say the southern). After crossing the river both take rest for 10 minutes and then start the return journey. During the return journey they cross each other at a distance of 400 m from the other bank (i.e. the northern). The two boats move at different but uniform speeds during the entire journey. How wide is the river?

94. Three friends starting at the same time from a point A reached simultaneously a point B which is 9 km away. They had 1 bicycle and 1 motor cycle with them. Only one person can ride a vehicle at one time, but the vehicle being used can be left anywhere for a friend to pick up and use. All the vehicles must also reach the destination when they reach. The uniform speeds of walking, bicycle and the motor cycle are, respectively, 5, 9 and 15 kmph. Determine the minimum time in which this journey can

be made satisfying all the conditions. Also mention how the vehicles are shared during the journey.

95. Two friends stood at the same point in the middle of a railway platform. When the engine of a moving train reached that location they started walking with same speed in opposite directions. The person walking in the opposite direction of the train crossed the end of the train after walking a distance of 20 m. The same thing happened with the other person (walking in the direction of the train) when he had walked 30 m. If the train was moving at a constant speed what is length of the train?

96. A person walks down slowly on a downward moving escalator. Using all the steps of the escalator he reaches the bottom in 50 steps. Then he walks up the same (downward moving) escalator taking steps five times faster and using all the steps of the escalator reaches the top in 125 steps. Again no step of the moving escalator was skipped. How many stairs of the escalator can be seen if it remains stationary?

97. A person went out of the house between 4 and 5 pm and returned between 5 and 6 pm; curiously he noticed that the hour and minute hands of the wall clock has interchanged their positions. How long did he spend outside?

98. Along a north-south highway four vehicles A, B, C and D are moving at different but uniform speeds. The first three are north bound whereas the fourth one is going in the opposite direction. Initially B was behind C and A was behind B while D was far north from all these three. A overtook B at 8 am and C at 9 am and was the first to meet D at 10 am. D met B at 12 noon and C at 2 pm. At what time did B overtake C?

99. In 1972 Ram was as old as the last two digits of his year or birth. The same is true for his grandfather. In what year were they born?

100. In the year x^2, a person was x years old. How old was he in the year 1950. No one lives more than 100 years.

101. The ages of Ram and his father both are two-digit numbers. If one writes first Ram's age followed by his father's then the resulting four-digit number is a perfect square. This fact was true even 25 years earlier. What are their current ages?

102. In this problem all numbers involved are natural numbers. A census officer noted down the number of a house before knocking on the door. The lady of the house opened the door. The officer asked "how many children do you have?" She answered three. The officer then asked the ages of the three children. She answered the product of their ages is 36. The officer, quite good at mathematics, could not figure out the individual ages and said so. The lady said "Ok, the sum of their ages is the number of this house". The officer looked at the number he has already noted and was still unable to

figure out the ages. Then the lady said my eldest son is a good singer. Then the officer noted down the individual ages of the children. What were the ages of the three children?

103. Chocolates are sold in two kinds of packets, one containing 3 and the other 10 chocolates. You can buy only complete packets. What is the maximum number of chocolates that you cannot buy?

104. Chocolates are sold in three kinds of packets, one containing 6, another 9 and the last one containing 20 chocolates. You can buy only complete packets. What is the maximum number of chocolates that you cannot buy?

105. Four married couples consumed 44 chocolates in total. Among the ladies Geeta took 2, Mita had 3, Neeta and Sita had, respectively 4 and 5 chocolates. It is known that Mr Acharjee took as many as his wife, Mr Banerjee had twice as many as his wife, Mr Chatterjee three times and Mr Mukherjee had four times as many as their respective wives. Determine the surname of these ladies and prove your answer.

106. 72 oranges cost ₹ __67.9__ , where the first digit of Rupees and the last digit of paise have been erased. What is the price of one orange?

107. Two brothers sold a herd of cows and the price of each cow in Rupees is exactly equal to the total number of cows they brought. When all the cows were sold they wanted to divide the money equally between them. Being somewhat weak in mathematics, they were unable to determine the share of each brother. So they decided to take by turn ₹ 10 from the total money earned. The elder brother was given the first turn. The process continued and at the end when the elder brother took his last turn, less than ₹ 10 was left for the younger brother. So to compensate the elder one gave his brother a new pen and by this transaction equal share was ensured. What is the price of the pen?

108. A bank teller, by genuine mistake, gave the customer money by interchanging the amount of Rupees and Paise written on the cheque. After spending 50 Paise from this amount, the customer noticed that she still has three times the amount written in the cheque. For what amount was the cheque.

109. A fruit vendor brought even and equal number of apples and oranges in his shop. He purchased each apple at a cost price of ₹ 1 and each orange at ₹ 2. He started selling all the fruits with a profit margin of 10 %. When seven fruits remain unsold he got back all his money he spent in buying these fruits. What will be his total profit when all the fruits are sold?

110. Grass grows at a uniform rate in a field which is already full of grass. A cow, a sheep and a rabbit all graze in this field. Each day the cow eats as much grass as the sheep and rabbit together consume per day. It is known that the cow and the sheep together make the field totally devoid of grass

in 45 days. The same job is accomplished by the cow and rabbit together in 60 days and the sheep and rabbit together in 90 days. How many days will be needed, to do this job, by all the animals together?

111. Unlimited supply of balls of four different colours, say, red, blue, yellow and green, is available. How many different types of packets, each containing 10 balls, can be made?

112. A large cake is to be distributed among 100 people. The first person will be served 1 % of the whole cake. The second person will be given 2 % of the remaining cake, the third person 3 % percent of the remaining cake and so on and finally the last person the entire remaining cake. Which position in the queue should be occupied to get the maximum amount of the cake?

113. The owner of a movie theatre declared that the birthday of every person in the queue will be noted. The first person having the same birthday as that of someone who has already purchased the ticket will be given a free entry. Which position in the queue should be occupied to maximize the chance of getting a free entry? Assume 365 days in a year.

114. A dishonest milkman removed 3 litres of milk from his full can and put 3 litres of water to replenish the volume. He repeated this process twice more. At the end of the third replenishment his milk became 50 % pure. Determine the capacity of the can in litres.

115. A cruel king came to know that one of the 1000 bottles of his very costly wine has been spiked with a deadly poison. The poison is such that even a drop of the spiked wine is sufficient to kill a man. He ordered his minister to identify the bottle for which a maximum of 10 convicts under death sentence can be sacrificed. But no one can be forced to drink twice. How can the minister carry out this order?

116. Solve Problem number 115 when the poison is such that the person having the spiked wine dies after 30 days and the minister has to accomplish the job within 31 days. All other conditions remain the same.

117. A toy company made a super-hard plastic egg which can be safely dropped from a height. They decided to determine the highest floor from which the egg can be dropped without breaking the egg. For this purpose a worker from the company took two identical eggs to a 100-storied building. Determine the minimum number of droppings required to definitely accomplish the goal so long the answer is less than or equal to 100th floor. Also explain how to carry out the job to find the answer.

118. In a deserted island five men and one monkey went to sleep with a heap of coconuts. In the middle of the night one man got up and divided the co-conuts in five equal parts and found one coconut extra, which he gave to the monkey. He took his one-fifth share and made one heap of the remain-

ing coconuts. After sometime another person got up and being unaware of the activity of the first person he did the same. He also took his one-fifth share of the remaining heap and found one extra coconut which he gave to the monkey. The process continued until the fifth person repeated the same activity. Next day morning when they got up they divided the remaining heap into five equal parts and nothing was left for the monkey. What is the minimum number of coconuts in the original heap before they went to sleep?

119. What will be the answer to Problem number 118 if on the next morning after final sharing between the men, the monkey again gets one coconut.

120. A person was shown 3 closed doors with the given information that behind two of these there is a goat each and in one of these there is a car. He was asked to choose one of the doors with the objective of getting the car. After he chose a particular door, another door was opened to show him a goat. Note that this can be always done independent of his choice. Now he was given a second chance to choose a door. Should he change his original choice to increase the probability of his getting the car? Explain your answer.

121. A person has to transfer 3000 bananas across a 1000 km desert. For this job he has only one camel which can carry a maximum of 1000 bananas. The man can only load and unload the camel with bananas as many times as he wants but he himself cannot carry any. Moreover, the camel must be given one banana per km, whether loaded or not. Bananas can be kept any-where on the desert for picking up later. Determine the maximum number of bananas that can reach the destination.

122. In the Fig. 4.1 an isosceles triangle with sides measuring 13, 13 and 10 is shown. A series of circles are drawn as indicated in this figure till the top vertex is reached. Find the sum of the perimeters of all these circles.

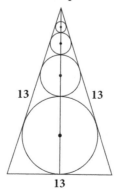

Fig. 4.1: Problem and solution No. 122

123. How many isosceles triangles with integer side-lengths exist if no side measures more than 1994? In this count equilateral triangles are considered as isosceles.

124. How many dissimilar triangles with integer angle measure can be drawn?

125. Consider an equilateral triangle of unit side (k=0, in Fig. 4.2). In the first step middle one-third of every side is erased and replaced by two line segments as indicated in Fig. 4.2 with $k = 1$. The same process is again repeated as shown in step $k = 2$ in Fig. 4.2. The process is repeated infinite times to get an imaginary curve which cannot be drawn. It approximately resembles a curve with a kink at every point. Determine the length of this imaginary closed curve and also the area circumscribed by it.

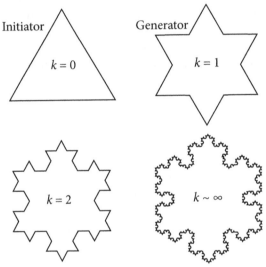

Fig. 4.2: Solution No. 125

126. In the following product of 100 terms which one term can be removed to render the product of the remaining 99 terms a perfect square?

$$N = (1!)(2!)(3!)\ldots(98!)(99!)(100!)$$

127. Partition the numbers 1 to 16 into two disjoint sets of equal size so that each set has the same sum, the same sum of the squares and the same sum of the cubes of their elements.

128. How many total positive integers less than 1000 can be written in binary notation using six or fewer 1's?

129. Prove that for any integer $n > 1$, the number $n^4 + 4^n$ cannot be a prime number.

130. (a) What is the first digit after the decimal point of $(\sqrt{2} + \sqrt{3})^{1000}$?

(b) Express the value of $\sqrt{13 + \sqrt{181 + 12\sqrt{5}}}$ as $m + \sqrt{n}$ with m and n as positive integers.

131. You first choose a three digit number in base 10. Then you convert this number to base 11 and base 16 and notice that these two new numbers have their digits in reverse order. What is the smallest number that you could have chosen?

132. Two persons A and B pick up 2 two-digit prime numbers. The square of A's number minus the square of B's number is equal to 20 more than the sum of eight times A's number and twelve times B's number. What is the greatest possible sum of A's and B's numbers?

133. Using each number at most once, Ram started with zero and added or subtracted some of the numbers from the list 1, 3, 9 and 27. How many different positive integers are possible for Ram to obtain?

134. You write the numbers 1, 2, 3, ..., 100 on a black board. Erase any two numbers m and n and write a new number $m + n + mn$. Note the total number on the board decreases by 1 in each step. What will be the last number left?

135. What is the sum of the digits of $\sqrt{(2021^2 - 4040) \times (2021^2 + 4044) - 4}$?

136. Start with the numbers 1 to 100 written on a blackboard. Erase any two numbers m and n and write a new number $\sqrt{m^2 + n^2}$. Continue the process until a single number is left on the board. What is that number?

137. Start with a pile of 100 pebbles. Split the pile into two (not necessarily equal) piles. Count the number in each of these two split piles and note the product of these two numbers. Go on splitting the piles, one at each step and note the product of the number of pebbles in the two freshly created piles in each step. Continue the process until you get 100 piles, each with one pebble. Determine the sum of all the products that have been noted.

Chapter 5

Solutions

1. Since $72 = 8 \times 9$, we conclude that the number is divisible by both 8 and 9. Divisibility by 8 implies that the three-digit number consisting of the last three digits of the given number, i.e. $73B$ is divisible by 8.

   Hence, $B = 6$. Thus, the number $A42736$ is divisible by 9, so the sum of the digits of this number, i.e. $A + 4 + 2 + 7 + 3 + 6 = A + 22$ is divisible by 9 which requires $A = 5$.

   Therefore, the number is 542736.

2. Let the erased two digits be x and y, respectively; i.e. the number is

$$24329020081xy640000.$$

This number must be divisible by both 9 and 11.

Divisibility by 9 implies the sum of the digits, i.e. $(41 + x + y)$ is divisible by 9. Since all digits are ≤ 9, it implies two possibilities for $(x + y)$, viz. $x + y = 4$ or 13. Since the number is divisible by 11 we take the difference of the sum of alternate digits as (notice that we start counting from the units place)

$$
\begin{aligned}
S_1 - S_2 &= (0 + 0 + 4 + y + 1 + 0 + 2 + 9 + +3 + 2) \\
&\quad - (0 + 0 + 6 + x + 8 + 0 + 0 + 2 + +4) \\
&= (21 + y) - (20 + x) \\
&= y + 1 - x.
\end{aligned}
$$

Again $y - x < 10$, so we conclude $x - y = 1$ so that $S_1 - S_2 = 0$ is divisible by 11 (0 is divisible by all numbers). Thus, we have to satisfy $x - y = 1$ and one of the other two equations obtained earlier, viz. $x + y = 4$ or $x + y = 13$. For integer values of x and y we have only one option $x + y = 13$. So finally we get $x = 7$ and $y = 6$.

3. From the indicated crosses it is easily concluded that the digit in the places of both tens and thousands of the quotient is 0. Moreover, the digit 8 in the place of hundreds in the quotient indicates that the divider is less than 125, since multiplication of the divider by 8 has resulted in a three-digit number. In the last step, four crosses indicate the product of the divider and the last digit (units place) is a four-digit number. So we conclude this last digit must be 9. So at this stage we can write the quotient as X0809. Now consider the first step of the division. Since we know the divider can be at most 124 and 124 × 7 = 868, which when subtracted from a four-digit number results in a three-digit number.

Therefore, we can conclude that the left most digit of the quotient must be 8 as the product of this digit with the divider is a three-digit number (hence < 9) and when this product is subtracted from a four digit number produces a two-digit number (hence > 7). Thus, we have completely determined the quotient as 80809. If we consider the divider as 123 we note that 123 × 80809 = 9939507, which is a 7 digit number. But the given dividend is a eight-digit number as evidenced by 8 crosses. So the divider must be 124.

Hence, we have determined both the divider and the quotient and hence the dividend as 124 × 80809 = 10020316 so the complete division sum can be written as

$$
\begin{array}{r}
124)10020316(80809 \\
992 \\
\hline
1003 \\
992 \\
\hline
1116 \\
1116 \\
\hline
\end{array}
$$

4. The sum of 3 three-digit numbers must be less than 3000. So A can be either 1 or 2. If we take A = 2, then considering the units place we conclude B = 8. In that case C must be 9 otherwise the sum will be less than 2000 and the assumption A = 2 would be wrong. With this solution, we get the problem as 289 + 892 + 928 = 2109 when in the places of tens and hundreds in the sum we are not getting 8(= B).

Hence, A is not 2. So we take A = 1. Again consider the units place, we get now, B = 9. Considering the place of tens (and the carryover from the units place) we conclude C = 8. Thus, we can verify the correct solution as 198 + 981 + 819 = 1998.

5. The sum of 2 four-digit numbers must be less than 20,000. So M = 1; when the second number is less than 2,000, i.e. the sum is less than 12,000. At this stage we can infer O is less than 2, i.e. O = 0 (as M = 1 has already been settled). Now we see that the second number MORE is less than 1,100 and

the sum is $\geq 10,100$, therefore, the first number SEND $\geq 9,000$, or, S = 9. For clarity, let us write the partial solution reached until now as

$$\begin{array}{r} 9\,E\,N\,D \\ +\,1\,0\,R\,E \\ \hline 1\,0\,N\,E\,Y \end{array}$$

Considering the digits in the place of hundreds it is easy to see that N = E + 1. Note that in this problem the carryover at any stage cannot be more than 1). Now remembering E < N, we can consider two possibilities by examining the digits in the tens place:

  (i) N + R = 10 + E (if D + E < 10) or,

  (ii) N + R + 1 = 10 + E (if D + E > 10, i.e. D + E = 10 + Y, note Y cannot be 0 as 0 has already been assigned to O). In other words there is a carryover 1 from the units place.

With the possibility (i), N + R = E + 1 + R = 10 + E implying R = 9, which cannot be true as 9 has already been assigned to S. So we proceed with the only possibility (ii) when R = 8.

Thus, we conclude for the letters (E, N, D and Y) six digits, namely, (2, 3, 4, 5, 6 and 7) are remaining. We know D + E = 10 + Y, so for the letters (Y, D and E), respectively, the following combinations are possible (2, 5 and 7), (2, 7 and 5), (3, 6 and 7), (3, 7 and 6). The third and fourth possibilities have to be discarded as discussed below. If Y = 3, D = 6 and E = 7, then N = E + 1 = 8, but 8 has already been assigned to R. Similarly, with Y = 3, D = 7 and E = 6, N = E + 1 = 7 which is being assumed to be D. Thus, the last possibility is also giving rise to a contradiction. For the same reason, the first possibility Y = 2, D = 5 and E = 7 has to be discarded, as again N = E + 1 = 8, which has already been assigned to R. Thus, finally we get (the second combination) Y = 2, D = 7 and E = 5, when N = E + 1 = 6.

Therefore, the final solution is obtained as S = 9, M = 1, O = 0, R = 8, Y = 2, D = 7, E = 5 and N = 6, i.e.

$$\begin{array}{r} 9\,5\,6\,7 \\ +\,1\,0\,8\,5 \\ \hline 1\,0\,6\,5\,2 \end{array}$$

6. First note that ten different letters, viz. F, O, R, T, Y, E, N, S, I, X have been used in this problem. So all the digits 0, 1, 2, 3, ..., 9 have been used up. Considering the places of units and tens and the absence or possibility of carryover, it is easy to conclude N = 0, E = 5. Moreover, from the places of hundreds and thousands (and the carryover) we write I = 1 and O = 9. Again considering the hundreds place and the carryovers we can write

$$2T + R + 1 = 20 + X \text{ or, } 2T + R = 19 + X \text{ and } S = F + 1$$

We have six digits, namely, (2, 3, 4, 6, 7, 8) to represent the undetermined six letters (F, R, S, T, X, Y). To satisfy the first of the above two equations we have the following three possibilities:

(i) T = 7, R = 8, X = 3; when with the remaining digits (2, 4, 6) available for (F, S and Y) we cannot satisfy the second equation S = F + 1 with any of the available choices.

(ii) T = 8, R = 6, X = 3; when again with the remaining digits (2, 4, 7) available for (F, S, Y) we cannot satisfy the second equation.

(iii) T = 8, R = 7, X = 4; with the remaining digits (2, 3, 6) available for (F, S, Y) we can satisfy the second equation with F = 2, S = 3 and hence Y = 6. Thus, the required solution demands N = 0, E = 5, I = 1, O = 9, T = 8, R = 7, X = 4, F = 2, S = 3 and Y = 6 and we can write the complete solution is given by

$$\begin{array}{r} 2\,9\,7\,8\,6 \\ 8\,5\,0 \\ +\,8\,5\,0 \\ \hline 3\,1\,4\,8\,6 \end{array}$$

7. Since the product consists of four digits, A ≤ 2, and A ≠ 0. Moreover, the place of units of the product has resulted by multiplying D by 4, so A must be even, and hence A = 2. Now multiplying D by 4 has resulted in 2, so the possible values of D is either 3 or 8. But considering the thousands digit in the product with A = 2, D cannot be 3 thus D = 8. Examining the hundreds and thousands we conclude B < 3 and considering the tens digits, we see B = 4 C + 1, and hence odd. So we get B = 1. since 4 C + 1 results in 1 the only option for C = 7. So the complete solution is

$$\begin{array}{r} 2\,1\,7\,8 \\ \times\,4 \\ \hline 8\,7\,1\,2 \end{array}$$

8. Since the product consists of only six digits considering the left most digit we conclude A = 1. Considering the digits in the units place (of the product, which is a multiple of 5), we conclude the digit E must be either 0 or 5. Let us consider these two possibilities separately.

(i) With E = 0.

As F ≥ 5 and E = 0, F can be either 6 or 8. First consider F = 6, when the digit in the place of tens suggests D = 3. Consequently examining the places of hundreds and thousands sequentially we find C = 5 and B = 6. But 6 has already been assigned to F. so this is wrong. Next we take F = 8, when following the product as before, we get D = 4 and then C = 0, but 0 has been assigned to E. So this is also wrong. Hence, we try the following option:

(ii) With E = 5.

The possible value of F now is either 7 or 9. First consider F = 9, when following the product as before D = 9, so an immediate contradiction (9 has been assigned to F). So we consider F = 7, when by carrying out the multiplication, sequentially we get D = 8, C = 2 and B = 4. Thus, the complete solution is

$$
\begin{array}{r}
1\,42\,8\,5\,7 \\
\times\,5 \\
\hline
7\,1\,42\,8\,5
\end{array}
$$

We may recall from eq. (3.8) where we showed that the rational fraction 1/7 in decimal notation is expressed as a recurring decimal in the form $\frac{1}{7}$ = $0.\dot{1}4285\dot{7}$ and multiplying this fraction by 2, 3, 4, 5, 6, we see the same set of digits appearing cyclically as a recurring decimal with a different starting digit and we can verify that $\frac{5}{7}$ = $0.\dot{7}1428\dot{5}$ as revealed in this problem.

9. First we note that neither A nor E can be 1. In the first case the product would have been ABCDE. In the second case A multiplied E would have given an A in the place of units in the product (and not E as given). Next we try E = 2 when E in the units place of the product says A must be 6 and the product is 222,222. But dividing this by 6 (i.e. A) results in ABCDE = 37037 which is wrong as digits are repeated for different letters.

Next we note E cannot be 3 as no value of A can result 3 (i.e. E) in the place of units in the product. If we take E = 4, then the only possible value of A, to result in 4 (i.e. E) in the place of units in the product, is 6 But the product 444,444 divided by 6 (i.e. A) results in ABCDE = 74074. Again we see the same mistake of repeating digits for different letters. Now take E = 5 when A = 3 or 7 or 9. with A = 3, dividing 555,555 by 3 (i.e. A), we get ABCDE = 555,555/3 = 185,185 again with repeating digits and with 6 digits!. So we take A = 7, when we get ABCDE = 555,555/7 = 79365 which gives back E = 5. So this is the correct answer as we write below:

$$
\begin{array}{r}
7\,9\,3\,6\,5 \\
\times\,7 \\
\hline
5\,5\,5\,5\,5\,5
\end{array}
$$

10. Let us write S = 98! + 99! + 100! = 98! (1 + 99 + 99 × 100) = 98!(100 + 99 × 100) = 98! (100 × 100) = 98! ×10^4.

So first we find the number of zeroes at the end of 98!. If we carry out the prime factorization of 98!, i.e. the product of all the numbers from 1 to 98, every time, we get a pair of 2 and 5, we get 10, which essentially means one zero is contributed at the end of this number. Since every alternate number is an even number, the number of 2's in this prime factorization will be many more than that of 5's. So the number of zeroes at the end will be the same as that of 5's. The number of 5's are given by $\lfloor \frac{98}{5} \rfloor + \lfloor \frac{98}{25} \rfloor$, where the sign $\lfloor x \rfloor$ denotes the floor function indicating the highest integer not greater than x. We include the second term as multiples of 25 (= 5^2) give two 5's. We stop at 5^2 as 5^3(= 125) is greater than 98. Thus, the number of zeroes at the end of the number S is 19 + 3 + 4 = 26. Note that the last 4 zeros are contributed by 10^4.

11. From the solution of Problem number 10, it can be seen that the number of zeroes at the end of the number $n!$ jumps whenever the value of n reaches an integral power of 5, say 5^p; the number of zeroes increases by p. Following this logic, we make the following Table showing the value of the number n and the number of zeroes at the end of $n!$

Number n	Number of zeroes at the end of $n!$
$5(= 5^1)$	1
$25(= 5^2)$	6
$125(= 5^3)$	31
$625(= 5^4)$	156
$3125(= 5^5)$	781
$15625(= 5^6)$	3906

It should be noted that in the above Table from the second row onward the entry in the second column is given by the sum of the two numbers in the previous row. From this table, we also conclude that the desired number in the problem, N lies between 3125 and 15625, as the number of zeroes (= 1993) in $N!$ lies between 781 and 3906. To get a close approximation of the number N, we write

$$\frac{N}{5} + \frac{N}{25} + \frac{N}{125} + \frac{N}{625} + \frac{N}{3125} \approx 1993 \text{ or, } \frac{718N}{3125} \approx 1993 \text{ or, } N \approx 7975$$

Now we find the number of zeroes at the end of 7995!, as

$$\left\lfloor \frac{7975}{5} \right\rfloor + \left\lfloor \frac{7975}{25} \right\rfloor + \left\lfloor \frac{7975}{125} \right\rfloor + \left\lfloor \frac{7975}{625} \right\rfloor + \left\lfloor \frac{7975}{3125} \right\rfloor = 1595 + 319 + 63 + 12 + 2$$
$$= 1991$$

It may be noted that in the above line, one needs to divide only by 5 as shown below:

$$\left\lfloor \frac{7975}{25} \right\rfloor = \left\lfloor \frac{1595}{5} \right\rfloor = 319; \quad \left\lfloor \frac{7975}{125} \right\rfloor = \left\lfloor \frac{319}{5} \right\rfloor = 63; \quad \left\lfloor \frac{7975}{625} \right\rfloor = \left\lfloor \frac{63}{5} \right\rfloor = 12$$

and so on.

Thus, to obtain the exact value of N, so as to get 1993 zeroes at the end of the factorial of that number, we need to go beyond 7975 to include two more multiples 5, viz. 7980 and 7985; also to be noted that none of these two is a multiple of 25.

Hence, the minimum value of N, to have 1993 zeroes at the end of $N!$, is $N = 7985$. The number of zeroes at the end of the factorial of a number does not change until by increasing the value a multiple of 5 is encountered.

Therefore, the possible 5 values of N are 7985, 7986, 7987, 7988 and 7989.

12. (a) The highest possible value of $n(= n_{max})$ is same as the number of 5's in the prime factorization of the number 100!. In other words, it is the number of zeroes at the end of 100!. So form the solution of Problem numbers 10 and 11, we easily obtain

$$n_{max} = \left\lfloor \frac{100}{5} \right\rfloor + \left\lfloor \frac{100}{25} \right\rfloor = 20 + \left\lfloor \frac{20}{5} \right\rfloor = 20 + 4 = 24.$$

(b) First we write the number 6 in terms of its prime factors as $6 = 2 \times 3$. We can easily see that in the prime factorization of 120!, the number of 2's will be more than the number of 3's. So the number of 6 will be decided by the number of 3's. The total number of 3's or the value of n_{max} is obtained as in part (a) as

$$n_{max} = \left\lfloor \frac{120}{3} \right\rfloor + \left\lfloor \frac{120}{9} \right\rfloor + \left\lfloor \frac{120}{27} \right\rfloor + \left\lfloor \frac{120}{81} \right\rfloor = 40 + 13 + 4 + 1 = 58.$$

13. $11^{10} - 1 = (10 + 1)^{10} - 1 = 10^{10} + 10 \times 10^9 + \cdots + 10 \times 10 + 1 - 1$.

So it is clearly seen after cancelling the last two 1's that all the terms are divisible 10^2. So the last two digits are 0's.

14. Expressing 12! in terms of prime factors we write

$$12! = 12 \times 11 \times 10 \times 9 \times 8 \times 7 \times 6 \times 5 \times 4 \times 3 \times 2 \times 1$$
$$= 2^2 \times 3 \times 11 \times 2 \times 5 \times 3 \times 3 \times 2^3 \times 7 \times 2 \times 3 \times 5 \times 2^2 \times 3 \times 2$$
$$= 2^{10} \times 3^5 \times 5^2 \times 7 \times 11$$
$$= (2^2)^5 \times (3^2)^2 \times (5^2)^1 \times 3 \times 7 \times 11$$

So considering the indices of only all the perfect squares, we get the total number of perfect square divisors is $(5+1) \times (2+1) \times (1+1) = 6 \times 3 \times 2 = 36$.

15. $N^2 = 25^{64} \times 64^{25} = (5^2)^{64} \times (8^2)^{25}$, therefore,

$$N = 5^{64} \times 8^{25} = 5^{64} \times 2^{75} = 10^{64} \times 2^{11} = 2048 \times 10^{64}$$

So the sum of the digits of N is $(2 + 0 + 4 + 8) = 14$ (Note all other digits are 64 number of 0's).

16. First we write both the numbers in terms of their respective prime factors as

$$7560 = 2^3 \times 3^3 \times 5^1 \times 7^1$$
$$8400 = 2^4 \times 3^1 \times 5^2 \times 7^1$$

To find the number of common divisors we consider the smaller of the two exponents of the same common prime factors and write the number of common divisors as

$$(3 + 1) \times (1 + 1) \times (1 + 1) \times (1 + 1) = 4 \times 2 \times 2 \times 2$$
$$= 32$$

17. First we write the number 24 in terms of its divisors in all possible ways as

$$24 = 24 \times 1 = 12 \times 2 = 8 \times 3 = 6 \times 4$$
$$= 6 \times 2 \times 2 = 4 \times 3 \times 2 = 3 \times 2 \times 2 \times 2$$

From these combinations the possible numbers having exactly 24 divisors can be written respectively as 2^{23}, $2^{11} \times 3^1$, $2^7 \times 3^2$, $2^5 \times 3^3$, $2^3 \times 3^2 \times 5^1$, $2^2 \times 3^1 \times 5^1 \times 7^1$. It can be seen that the smallest possible number is given by the second last number, i.e. $8 \times 9 \times 5 = 360$.

18. First we write the number in terms of its prime factors as $2160 = 2^4 \times 3^3 \times 5^1$. Hence, the total number of divisors is

$$(4 + 1) \times (3 + 1) \times (1 + 1) = 5 \times 4 \times 2 = 40.$$

Now we determine the number of odd divisors as $(3+1) \times (1+1) = 4 \times 2 = 8$. So the number of even divisors is $40 - 8 = 32$.

Alternatively, for an even divisor there must be at least a prime factor 2, so we can directly get that number as

$$(4 + 0) \times (3 + 1) \times (1 + 1) = 4 \times 4 \times 2 - 32.$$

19. Again we start with the prime factorization of 2160 as $2160 = 2^4 \times 3^3 \times 5^1$. So the number of divisors divisible by 3 is

$$(4 + 1) \times (3 + 0) \times (1 + 1) = 30$$

20. (a) Let the prime factorization of the number $N = p_i^{n_i} \times p_j^{n_j}$; where p_i and p_j are the two prime factors. It is given that $(n_i + 1)(n_j + 1) = 9$, so we get $n_i = 2$, $n_j = 2$. Thus, we get $N = p_i^2 \times p_j^2$; hence $N^2 = p_i^4 \times p_j^4$. The number of divisors of N^2 is given by $(4 + 1) \times (4 + 1) = 25$.

(b) Let the number be of the form M^3, then the prime factorization of M must have indices of all prime factors as multiples of 3. Next we note that for the total number of divisors to be 28, there are two ways of expressing it with indices of prime factors as multiples of 3

(i) $28 = 4 \times 7 = (3+1)(6+1)$ (ii) $28 = (27+1)$.

So for (i) we can write $M = p_1 \times p_2^2$ and for (ii) $M = p_3^9$, where p_i's indicate primes. Thus, the number of divisors of $M = (1+1) \times (2+1) = 6$ for (i) and $(9+1) = 10$ for (ii).

Hence, the total possible number of divisors for all possible values of M is $6 + 10 = 16$.

(c) The number 100 can be factorized many ways such as

   (i) 100×1 (iii) 25×4

   (ii) 50×2 (iv) $25 \times 2 \times 2, \ldots, 5 \times 5 \times 2 \times 2$.

For the desired number N to have exactly 100 factors we express the number in terms of its prime factors as $N = 2^{n_1} \times 3^{n_2} \times 5^{n_3} \times 7^{n_s} \times \ldots$, where for the different factorizations given above, to determine the smallest value of N, we get

   (i) $N = 2^{99}$ (iii) $N = 2^{24} \times 3^3, \ldots.$

   (ii) $N = 2^{49} \times 3$

Continuing this way and comparing different values of N, we find finally the smallest value of N from the last set of factors of 100 mentioned above as $N_{\min} = 2^4 \times 3^4 \times 5 \times 7 = 45{,}360$.

21. (a) First it must be noted that the square of any prime number p, i.e. p^2 has exactly three divisors, viz. 1, p and p^2. So we write $N_1 = p_i^2$, $N_2 = p_j^2$ and $N_3 = p_k^2$.

Thus, we get $N_1^2 N_2^3 N_3^4 = p_i^4 p_j^6 p_k^8$.

Therefore, the number of divisors of this product is

$$(4+1) \times (6+1) \times (8+1) = 5 \times 7 \times 9 = 315.$$

(b) Let the bigger and the smaller diameters be given by two integers m and n, with $m > n$. It is given that $\frac{\pi}{4}(m^2 - n^2) = 36\pi$ or, $(m^2 - n^2) = 144$. So $(m+n)(m-n) = 144$. We can factorize 144, in terms of two unequal factors in different ways, such as 144×1, 72×2, 48×3, 36×4, 24×6, 18×8, 16×9. With integers m and n, both $(m+n)$ and $(m-n)$ must be either both odd or both even. So considering only these possibilities, equating these two values with the corresponding factors, we get the following values of $n = 35$, 16, 9 and 5. The sum of all these possible values is $35 + 16 + 9 + 5 = 65$.

(c) Let the two integers be m and n, with $m > n$. Given $\frac{1}{m} + \frac{1}{n} = \frac{1}{12}$ or, $mn - 12(m+n) = 0$.

Hence,

$$mn - 12m - 12n + 144 = 144$$

$$\text{or, } (m-12)(n-12) = 144$$

Using the factors of 144, used in part (b), we get n = 13, 14, 15, 16, 18, 20, 21. Thus, the sum of all the possible values of the smaller number is 13 + 14 + 15 + 16 + 18 + 20 + 21 = 117.

(d) The first equation is written as

$$ab - 4a - 2b = 34$$
$$\text{or, } a(b - 4) - 2(b - 4) = 42$$
$$\text{or, } (b - 4)(a - 2) = 42$$

Equating $(b - 4)$ with various divisors of 42, i.e. 1, 2, 3, 6, 7, 14, 21 and 42, we get the possible values of b as 5, 6, 7, 10, 11, 18, 25 and 46.

Similarly, the second equation can be factored as $(b - 3)(c - 5) = 49$. For $(b - 3)$ to be a divisor of 49, i.e. 1, 7 and 49, possible values of b are, respectively 4, 10, 52. To satisfy both the equations, the only possible value of b = 10, when the values of a and c turn out to be 9 and 12, respectively. So the value of $a + b + c = 9 + 10 + 12 = 31$.

22. $\dfrac{n^3 - 3}{n - 3} = \dfrac{n^3 - 27 + 24}{n - 3} = \dfrac{n^3 - 3^3}{n - 3} + \dfrac{24}{n - 3} = \left(n^2 + 3n + 9\right) + \dfrac{24}{n - 3}.$

Now we know the divisors of 24 are 1, 2, 3, 4, 6, 8, 12, 24, therefore, the required divisibility will be satisfied for n = 1, 2, 4, 5, 6, 7, 9, 11, 15, 27 when $n - 3$ = −2, −1, 1, 2, 3, 4, 6, 8, 12, 24. Note the first two values of n, which makes the divisor of 24 as negative integers −2 and −1 respectively.

23. First we may note the following pattern in the products considered:

$$33 \times 33 = 1089$$
$$333 \times 333 = 110889$$
$$3333 \times 3333 = 11108889$$

and convince ourselves that the answer to the given problem will be 332 number of 1's followed by 1 zero, then 332 number of 8's and ending with a 9. Without such a guess we can now solve the problem mathematically as written below:

$$S^2 = (1/3)(999\ldots999)(1/3)(999\ldots999)$$
$$= (1/9)(10^{333} - 1)^2$$
$$= (1/9)(10^{666} - 2 \times 10^{333} + 1)$$
$$= (1/9)(999\ldots98000\ldots01)$$
$$= (111\ldots1110888\ldots8889)$$

The reader should determine how many 1's and how many 8's are there in the above expression.

24. (a) $9S = 10-1+10^2-1+10^3-1+\cdots+10^{111}-1 = 111\ldots1110-111$, where in the first term the digit 1 appears 111 times $= 111\ldots1111 - 112$, where in the first term the digit 1 has appeared 112 times.

    Or, $S = (111\ldots1111 - 112)/9$. Now let us note the result of dividing a number consisting of nine 1's by 9, i.e. 111, 111, 111/9 = 12345679. Thus, all the digits 1 to 9 appear serially, except the digit 8, in the quotient.

    Therefore, in S, we will see 12345679 repeated side by side 12 times and the remaining part will be $(1111 - 112)/9 = 0111$. The sum of the digits of S can now be easily calculated as

    $$12(1 + 2 + 3 + 4 + 5 + 6 + 7 + 9) + 1 + 1 + 1 = 12 \times 37 + 3$$
    $$= 447$$

    (b) Note that all three-digit palindromes can be written as $101m+10n$, with integers m and n in the range $1 \le m \le 9$ and $0 \le n \le 9$. So the sum of all three-digit palindromes is

    $$10(101 + 202 + 303 + \cdots + 909) + 9(10 + 20 + 30 + \cdots + 90)$$
    $$= 45{,}450 + 4050$$
    $$= 49500$$

    (c) Note that all three-digit palindromes can be written as $101m+10n$, with integers m and n in the range $1 \le m \le 9$ and $0 \le n \le 9$. Using the divisibility rule for 11, we conclude $2m-n$ is either 0 or a multiple of 11 for the three-digit palindrome to be divisible by 11. These conditions are satisfied for the ordered pair (m, n) given by $(1, 2), (2, 4), (3, 6), (4, 8),$ $(6, 1),(7, 3), (8, 5)$ and $(9, 7)$. Thus, there are 8 three-digit palindromes which are divisible by 11. In total we have 9 (possible values of m) $\times$ 10 (possible values of n) = 90 three-digit palindromes.

    Hence, there are $90 - 8 = 82$ three-digit palindromes which are not divisible by 11.

25. Let the number be $N = abccba$; $a \neq 0$. To judge the divisibility by 7 we rewrite the number as

$$N = abccba = (10^5 + 1)a + (10^4 + 10)b + (10^3 + 10^2)c$$
$$= (7 \times 14285 + 6a) + (7 \times 1430)b + (7 \times 157 + 1)c$$

Thus, for 7 to divide N, the number $(6a + c)$ must be divisible by 7. For this to happen, the following combinations of a and c are acceptable:

(i) $a = c = 1, 2, 3, 4, 5, 6, 7, 8, 9$ (iv) $a = 2, c = 9$

(ii) $a = 1, c = 8$ (v) $a = 9, c = 2$

(iii) $a = 8, c = 1$ (vi) $a = 7, c = 0$.

Hence, there exist $9 + 5 = 14$ combinations of a and c. For each of these 14 combinations, the digit b can take any of the five odd digits, viz. 1, 3, 5, 7, 9. So the total number of possible N satisfying all the conditions is $14 \times 5 = 70$.

26. (a) We know that $(a^n - b^n)$ is divisible by $(a - b)$ for all integer values of n and $(a^n + b^n)$ is divisible by $(a + b)$ only for odd values of n.

Moreover, we note that $(5555 - 4 = 5551)$ and $(2222 + 4 = 2226)$ both are divisible by 7. With these two observations, now we write the given expression

$$5555^{2222} + 2222^{5555}$$
$$= 5555^{2222} - 4^{2222} + 2222^{5555} + 4^{5555} + 4^{2222} - 4^{5555}$$
$$= 5551x + 2226y - 4^{2222}(4^{3333} - 1)$$

where x and y are two integers. We can continue to write the above number as

$$= 5551x + 2226y - 4^{2222}(4^{3 \times 1111} - 1)$$
$$= 5551x + 2226y - 4^{2222}(64^{1111} - 1)$$
$$= 5551x + 2226y - 4^{2222} \times 63z$$

where z is another integer. Note that as all the terms in the above expression are divisible by 7, the whole number is also divisible by 7.

(b) We write $65^{2021} = (63 + 2)^{2021}$. Expanding this binomial expression we can see all the terms are divisible by 63 except 2^{2021}. Now we write $2^{2021} = (2^6)^{336} \times 2^5 = (63 + 1)^{336} \times 32$. Note that the first term is written using binomial expansion, we get $(63 + 1)^{336} \times 32 = (K.63 + 1) \times 32$, where K is a positive whole number. Thus, when this number is divided by 63, the remainder is 32.

27. $N = 6 + 66 + 666 + \cdots + \underbrace{666\ldots666}_{100}$ or

$$N = \tfrac{6}{9}[9 + 99 + 999 + \cdots + \underbrace{999\ldots999}_{100}]$$

$$= \tfrac{6}{9}[(10 - 1) + (10^2 - 1) + (10^3 - 1) + \underbrace{\ldots}_{100} + (10^{100} - 1)]$$

$$= \tfrac{6}{9}[\underbrace{111\ldots1110}_{100} - 100]$$

$$= \tfrac{6}{9}[\underbrace{111\ldots111}_{99}\, 10 - 100]$$

Now we need to note that $\frac{111,111,111}{9} = 12345679$, i.e. all the digits from 1 to 9, except 8, appear sequentially. Using this observation, we can write

$$N = 6[\underbrace{123456790123456790\ldots123456790}_{10}\quad 1234567890]$$

$$= [\underbrace{740740740740740740\ldots740740740}_{10}\quad 7407407340]$$

Thus, in place value notation the number N is written by writing 740 thirty two times side by side followed by 7340. Thus, the digit 7 will appear 33 times.

28. This problem will be solved by using the method of induction. Towards this end, first we write $f(n) = n^5 - n$ and calculate $f(1) = 0$, $f(2) = 30$, $f(3) = 240$, $f(4) = 1020$ and $f(5) = 3120$. Note that all these are divisible by 5. next we show that if $f(n)$ is divisible, then so is $f(n + 5)$. first we write

$$f(n + 5) = (n + 5)^5 - (n + 5)$$
$$= n^5 + 5n^4 \cdot 5 + 10n^3 \cdot 5^2 + 10n^2 \cdot 5^3 + 5n \cdot 5^4 + 5^5 - (n + 5)$$
$$= f(n) + 25n^4 + 250n^3 + 1250n^2 + 3125n + 3120$$

So if $f(n)$ is divisible by 5, then so is $f(n + 5)$ as all other terms on the righthand side of the above equation are divisible by 5.

Hence, we conclude by method of induction, that if $f(1)$, then so is $f(6)$. This argument can be extended in steps of five to cover all the values of n.

Alternative solution:

$$n^5 - n = n(n^4 - 1) = n(n^2 - 1)(n^2 + 1)$$
$$= n(n - 1)(n + 1)(n^2 - 4) + 5]$$
$$= (n - 1)n(n + 1)(n - 2)(n + 2) + 5n(n - 1)(n + 1)$$
$$= (n - 2)(n - 1)n(n + 1)(n + 2) + 5n(n - 1)(n + 1)$$

It is readily seen that the first term is a product of 5 consecutive positive integers (for $n > 2$), hence one of them is divisible by 5. For $n = 1$ and 2, one of the five integers in the first term is zero, and so the first term will be 0 and 0 is divisible by all numbers including 5. Also the second term is a multiple of 5.

Hence, $(n^5 - n)$ is divisible by 5 for all values of n.

29. (a) We write the product of four consecutive numbers as

$$N = n(n + 1)(n + 2)(n + 3)$$
$$= (n^2 + 3n)(n^2 + 3n + 2)$$
$$= [(n^2 + 3n + 1) - 1][(n^2 + 3n + 1) + 1]$$
$$= (n^2 + 3n + 1)^2 - 1$$

Thus, this number is seen to be 1 less than a perfect square.

(b) Put $n = 2019$, then the given number is $\sqrt{1 + n(n + 1)(n + 2)(n + 3)}$. Using the result of part (a), we can write this number as $(n^2 + 3n + 1)$, which is $(2019^2 + 3 \times 2019 + 1) = 4082419$. The sum of the digits is 28.

30. If a perfect square number is even, then it can be expressed as $(2m)^2 = 4m^2$, hence divisible by 4. If a perfect square number is odd, then it can be written

$$(2m + 1)^2 = 4m^2 + 4m + 1 = 4m(m + 1) + 1$$

Now one of the two consecutive numbers, viz. m and $(m + 1)$ must be even and has a factor 2, so the square number is $8x + 1$, where x is an integer. Thus, the number, when divided by 8, leaves a remainder 1.

31. Let

$$2m - n = p^2 \tag{i}$$
$$m - 2n = q^2 \tag{ii}$$
$$m + n = r^2 \tag{iii}$$

Subtracting (ii) from (i), we get

$$m + n = p^2 - q^2 \tag{iv}$$

From (iii) and (iv), it is seen

$$p^2 = q^2 + r^2 \tag{v}$$

Thus, (p, q, r) is a Pythagorean triple. It is also easy to see that $p > r > q$. Again subtracting (ii) from (iii), we get

$$3n = r^2 - q^2 \tag{vi}$$

From (vi) it is not difficult to conclude all the three numbers, viz. n, r, q are divisible by 3.

Hence, from (v), p is also divisible by 3. We know that the smallest Pythagorean triple is (3, 4, 5).

Hence, the smallest possible values of (q, r and p), are respectively, (9, 12 and 15). So from (vi), the smallest possible value of n is $(144 - 81)/3 = 21$.

32. All prime numbers greater than 3 can be written as $(6n \pm 1)$; where n is an integer. Note that the converse is not true, i.e. all numbers of these forms are not prime. It may be noted that $(6n + 2)$ and $(6n + 4)$ are even numbers (greater than 2) and $(6n + 3)$ must be divisible by 3, hence all these numbers are composite.

So we write all such primes as $p = 6n+1$, or, $p^2 - 1 = 36n^2 \pm 12n = 12n(3n \pm 1)$.

Not one of the numbers n or $(3n \pm 1)$ must be even, so will have a factor 2, hence $(p^2 - 1)$ must be divisible by 24.

Alternatively, from problem 29(a), we conclude that the given number is a product of four consecutive numbers. So one of these is a multiple of 2, one a multiple of 3 and another a multiple of 4. So the product of all these will be a multiple of 24. Thus, the number is divisible by 24.

33. We can easily write

$$m^2 - 2m = m(m - 2) = n(n - 1)(n^2 - n - 2)$$
$$= n(n - 1)(n - 2)(n + 1)$$
$$= (n - 2)(n - 1)(n)(n + 1)$$

Thus, the expression is a product of four consecutive numbers. Consequently one of them must be a multiple of 3 and the two of them will be two consecutive even numbers and hence one is a multiple of 2 and the other a multiple of 4. So the expression is a multiple of $2 \times 3 \times 4$, i.e. 24.

34. The product of the digits $P(n)$ of any number n is greater than or equal to zero. So

$$P(n) = n^2 - 15n - 27 \geq 0$$
$$\text{or, } n \geq \frac{15 \pm \sqrt{225 + 108}}{2}, \text{ i.e. } n \geq 16.60\ldots$$

Note n is positive. We will now show that $n \geq P(n)$, where the equality sign is applicable for only single digit numbers. For an m-digit number n, we write

$$n = a_{m-1}10^{m-1} + a_{m-2}10^{m-2} + \cdots + a_1 10^1 + a_0$$
$$\geq a_{m-1}9^{m-1}$$

If all the digits except the first one takes the maximum value 9, then $P(n) = a_{m-1}9^{m-1}$. Thus, we can write $n \geq P(n)$, where the equality sign hold only for $m = 1$. Hence, we get

$$P(n) = n^2 - 15n - 27 \leq n \text{ or, } n^2 - 16n - 27 \leq 0$$

$$\text{or, } n \leq \frac{16 \pm \sqrt{256 + 108}}{2}$$

Considering only the positive value, we write $n \leq 17.50\dots$.

From (i) and (ii), $n = 17$.

35. To solve this problem we use the solution to Problem number 30, where we have seen that a perfect square number is either divisible by 4 or leaves a remainder 1 when divided by 8. from the divisibility rule we conclude none of the numbers is divisible by 4 as 11 is not divisible by 4. Also since 111 does not leave a remainder 1 when divided by 8, none of the numbers does so. hence none of these numbers can be a perfect square.

36. $n^4 + 4n^2 + 11 = n^4 + 4n^2 - 5 + 16 = (n^2 + 5)(n^2 - 1) + 16$.

Now in Problem 30, we have seen that for odd values of n, $(n^2 - 1)$ is divisible by 8 and we note that for odd values of n, $(n^2 + 5)$ is even and hence definitely has a factor 2, so the given number is obviously divisible by 16.

37. Substituting for r from the first equation into the second, we get

$$p^2 + q^2 - r^2 = p^2 + q^2 - (p + q - 1)^2 = -1$$
$$\text{or, } p^2 + q^2 - p^2 - q^2 - 2pq - 1 + 2(p + q) = -1$$
$$\text{or, } p + q - pq = 0$$
$$\text{or, } (p - 1)(1 - q) = -1$$

Thus, we get either, $p - 1 = 1$, $1 - q = -1$, i.e. $p = 2$, $q = 2$, or, $p - 1 = -1$, $1 - q = 1$, i.e. $p = 0$, $q = 0$.

From the first equation given in the question, we get the corresponding values of $r = p + q - 1 = 3$ and -1, respectively.

Thus, the sum of all possible values of

$$p^2 + q^2 + r^2 = (2^2 + 2^2 + 3^2) + (0^2 + 0^2 + (-1)^2) = 17 + 1 = 18$$

38. Let $n = p^2 + q^2$ when $2n = 2(p^2 + q^2) = (p + q)^2 + (p - q)^2$, i.e. sum of two perfect squares.

Conversely if $2n = a^2 + b^2$, then the numbers a and b are both even or both odd. Now it can be readily seen that

$$n = \left(\frac{a + b}{2}\right)^2 + \left(\frac{a - b}{2}\right)^2$$

where both the parts in the righthand side are integers because a and b are either both even or both odd. So the statement and its converse are proved.

39. Let $n^2 + 19n + 47 = m^2$ or, $4(n^2 + 19n + 47) = (2m)^2$.

We multiplied by 4 to avoid fractions in the subsequent step. Now we write from the above equation in the form

$$(2n + 19)^2 - 173 = (2m)^2$$
$$\text{or, } (2n + 19)^2 - (2m)^2 = 173$$
$$\text{or, } (2n + 19 + 2m)(2n + 19 - 2m) = 173$$

Now we note that 173 is a prime number so it can only be written as one of the four ways to give us the value of the factors on the lefthand side of the above equation:

(i) 1×173 (iii) -1×-173

(ii) 173×1 (iv) $-173x - 1$.

Now considering all the four possibilities we finally get $n = 34$ and -53 with $m = 43$ or -43.

40. In each of these numbers the sum of the digits is $4 \times 4 + 3 \times 3 + 2 \times 2 + 1 = 30$. So this sum and hence the number is divisible by 3 and not by 9. So in prime factorization a single 3 will be available and hence this is not a perfect square as that requires all prime factors must appear even times.

41. The lowest three sums must include the lowest number p and the next three numbers. Also the highest five sums must include the highest number, u and sequentially the other five numbers. From these conditions we can conclude that the following equations to hold good:

$$p + q = 25 \tag{i}$$
$$p + r = 30 \tag{ii}$$
$$p + s = 38 \tag{iii}$$
$$p + u = 79 \tag{iv}$$
$$q + u = 90 \tag{v}$$

Subtracting (iv) from (v), we get

$$q - p = 11 \tag{vi}$$

Subtracting (vi) from (i), we get $2p = 14$ or, $p = 7$. Now using (ii) and (iii), it is easy to see $r = 23$ and $s = 31$. So $r + s = 54$.

42. The smallest number which is divisible 5, 7 and 8 is the LCM (lowest common multiple) of these three numbers which is easily seen to be $5 \times 7 \times 8 =$

280 (as three numbers are co-primes or relatively prime, i.e. have no common factor).

Hence, the required smallest number is 280 + 1 = 281.

43. Notice that in if the required number is increased by three then it becomes the lowest multiple of 5, 8 and 11. So the required number is LCM (5, 8 and 11) −3, i.e. 440 − 3 = 437.

44. (a) Let the common remainder be r. So the required number is the gcd (greatest common divisor, also known as the *highest common factor* HCF) of $(364 - r)$, $(414 - r)$ and $(539 - r)$. Let the gcd be k. Then we can write

$$364 - r = ka \qquad\qquad (i)$$
$$414 - r = kb \qquad\qquad (ii)$$
$$539 - r = kc \qquad\qquad (iii)$$

where a, b and c do not have any common divisor.

Subtracting (ii) from (iii), (i) from (iii) and (i) from (ii), we get $125 = k(c - b)$, $175 = k(c - a)$ and $50 = k(b - a)$.

Since $(c - b)$, $(c - a)$ and $(b - a)$ cannot have any common factor, k is the gcd of $(125, 175, 50)$. Now $125 = 25 \times 5$, $175 = 25 \times 7$ and $50 = 25 \times 2$. As 5, 7 and 2 do not have any common factor, we get $k = 25$. (One can also use the common Euclid's algorithm to find the gcd of 50 and 125 and then that of this gcd and 175. The reader may see the solution of Problem number 108).

(b) We can find 10 is the smallest number which satisfies all the conditions. Now LCM of $(6, 7, 8)$ is 168. So all numbers of the form $168m + 10$, with m as a positive integer, will satisfy the required conditions. Now for the number to be three-digit number, the maximum value of m is 5. Hence, the number is $5 \times 168 + 10 = 850$.

(c) Let the number be n, then the following equations must be satisfied:

$$n = 23x + 2 \qquad\qquad (i)$$
$$n = 232y + 23 \qquad\qquad (ii)$$
$$n = 2323z + 232 \qquad\qquad (iii)$$

where x, y and z are integers and $x > y > z$. To determine the minimum value of the number n, we first note from (i) and (iii) that $23x + 2 = 2323z + 232$ or, $x = 101z + 10$.

So an integer value of z automatically results in an integer value of x. From (ii) and (iii), we can write

$$y = 10z + \frac{3z + 209}{232}$$

Thus, to get an integer (minimum) value of y, we make the numerator of the second term on the righthand side of the above equation the smallest possible integer which results in an integer value of z.

Hence, we set $3z + 209 = 2 \times 264$, when $z = 85$.

Note that $3z + 209 = 1 \times 264$ results in a non-integer value of z. With $z = 85$, we get using (iii), the minimum value of

$$n = 2323 \times 85 + 232 = 197{,}687$$

45. For the number to be odd, the last digit must be odd, i.e. anyone out of 1, 3, 5, 7, 9. So there are 5 choices for the units digit. Now we consider the thousands digit. We have 8 choices for this digit, as we cannot use 0 and the digit that has already been used in the unit's place. For the digits of the remaining two places of the four-digit number we have 8 and 7 choices, respectively. Note that here the digit 0 is allowed. So the total number of numbers satisfying all the conditions is $5 \times 8 \times 8 \times 7 = 2{,}240$.

    Alternative solution: We have 5 choices for the units digit. Then we have 9, 8 and 7 choices, respectively for the other three places. So the number of different numbers is $5 \times 9 \times 8 \times 7 = 2{,}520$. But now, we have to subtract the number of those which starts with 0. This number is $5 \times 1 \times 8 \times 7 = 280$. So the number of four-digit odd numbers with distinct digits is $2{,}520 - 280 = 2{,}240$.

46. The units digit must be either 0 or 5. We consider these two possibilities separately as follows:

    (i) We put 0 in the unit's place. Then the greatest digit 5 can be placed in three different locations. Now to fill up the remaining two places we are left with four digits, viz. 1, 2, 3 and 4. This can be done in $4 \times 3 = 12$ different ways. So the number of this kind of numbers is $3 \times 12 = 36$.

    (ii) We put 5 in the unit's place. So we have four choices for the thousand's place, viz. 1, 2, 3 and 4. Now we are left with 4 digits (including 0) to put in the remaining two places. This can be done in $4 \times 3 = 12$ different ways. So the number of this kind of numbers is $4 \times 4 \times 3 = 48$.

    Thus, the answer to the given problem is $36 + 48 = 84$.

    Note that all the restrictions are satisfied before the free choices are considered.

47. If we write sequentially all the numbers from 00000 to 99999, then we get at least one digit different in all these numbers. Now to make at least two digits different we add a sixth digit (i.e. create a new units digit). If two digits were different in the already written 10^5 numbers, from 00000 through 99999, then any digit can be in the newly created unit's place, we need not

bother. So we put here the digit that occurs in the unit's place of the sum of the digits of the five-digit number that has been written. This digit will definitely be different for all these five-digit numbers whose digits originally differed only in one place. Note that the difference in these sums of the digits cannot be 0 or 10. Thus, we get 10^5 numbers whose digits differ at least in two places. Thus, we can create at most 10^5 identity cards with numbers whose digits differ at least in two places and we have also explained how to achieve this.

48. First let us pair up all the numbers from 0 to 99,999 as written below within a parenthesis.

    $(0, 99999), (1, 99998), (2, 99997), \ldots (49997, 50002), (49998, 50001)$ and $(49999, 50000)$.

    Note that the sum of the digits of all these pairs is $9 \times 5 = 45$. The number of such pairs is 50,000. Thus, the sum of the digits of all the numbers from 0 through 99,999 is $45 \times 50,000 = 225 \times 10^4$.

    Now we add the digit of the number 100,000 which is 1.

    Hence, the answer to the problem is $225 \times 10^4 + 1$.

    Alternative solution: When we write 00000 to 99999 we write 10^5 five-digit numbers. So we write in all 5×10^5 digits. There is no preference between the different digits, all ten digits 0, 1, 2, ..., 9 are written equal number of times. Thus, each digit is written 5×10^4 times. So the sum of the digits of these 10^5 numbers is $5 \times 10^4 \times (0 + 1 + 2 + 3 + 4 + 5 + 6 + 7 + 8 + 9) = 5 \times 10^4 \times 45 = 225 \times 10^4$.

    Hence, the answer to the problem by considering the number 100,000 is $225 \times 10^4 + 1$.

49. We know that there are 90,000 five-digit numbers. Out of these, every third number is a multiple of 3, thus the number of five-digit multiples of 3 is 30,000. First we determine how many of these do not contain the digit 6. Then by complimentary counting, i.e. by subtracting that from 30,000, we get the five-digit multiples of 3 which contain at least one 6. For the first (from left) digit we have eight choices, all the digits except 0 and 6. Then for the next three places we have 9 choices, all the other digits except 6. Now the sum of the digits in these first four places, when divided by 3 there are three possible remainders (i) 0, (ii) 1 and (iii) 2. For the five-digit number to be a multiple of 3 the sum of all the five digits must be divisible by 3. Thus, in case (ii) we can have 2, 5 or 8 and in case (iii) again three choices, viz. 1, 4 or 7. Thus, we have 3 choices for the units digit of all the possible combinations.

Hence, the number of five-digit multiples of 3 without the digit 6 appearing anywhere is $8 \times 9 \times 9 \times 9 \times 3 = 17{,}496$. So the number of such multiples of 3 with at least one 6 is $(30{,}000 - 17{,}496) = 12{,}504$.

50. In this problem again we use complimentary counting. First we see 10^6 numbers, from 000000 to 999999, when the counter runs through its entire range. If the digit 1 is not used in any of the windows then we can see 9^6 numbers as all the other 9 digits appear in every window. Thus, the digit 1 is displayed at least once in $(10^6 - 9^6) = 468{,}559$ of the numbers that are displayed in this counter.

    The total number of digits displayed as the counter runs through its entire range 6×10^6 as 10^6 numbers each of six-digit are displayed. All ten digits are displayed equal number of times as there is no preference for a particular digit. So each digit, like 1, is displayed $0.1 \times 6 \times 10^6 = 6 \times 10^5$ times.

51. Let the seven numbers be $a < b < c < d < e < f < g$, then if $e+f+g < 50$, then $a+b+c+d > 50$, i.e. $d > 14$, because with $d = 14$, we get the maximum value of $a+b+c+d = 11+12+13+14 = 50$. Thus, in that case the minimum value of $d = 15$, when the minimum value of $e + f + g = 16 + 17 + 18 = 51$, i.e. > 50. So we are landing in a contradiction.

    Hence, the minimum value of $e + f + g$ is at least 50.

52. Consider the following numbers: $S_1 = a_1$, $S_2 = a_1 + a_2$, $S_3 = a_1 + a_2 + a_3$, $\ldots$, $S_n = \sum_{i=1}^{n} a_i$.

    If any of these S_i's, $1 \leq i \leq n$ is divisible by n, then the proposition is immediately proved. If not, then the n values of the remainders all cannot be different as the remainder can have at most $(n-1)$ different values, viz. $1, 2, 3, \ldots, (n-1)$. Thus, at least two (in the worst case) of the remainders must be same (Pigeon Hole Principle — if n pigeons are to be placed in $(n-1)$ cages, then at least one cage must contain at least two pigeons). Let those two values of S_i's are indicated by S_k and S_l with $1 \leq k, l \leq n$ and $l > k$. We can definitely say then that $(S_l - S_k)$ is divisible by n.

    Hence, the sum of the elements of the subset $\{a_{k+1}, \ldots, a_l\}$ is divisible by n.

53. You first ask your friend the value of the polynomial at $x = 1$ and say you get the answer as N.

    Thus,

    $$P(1) = \sum_{i=0}^{n} a_i = a_0 + a_1 + a_2 + \cdots + a_n = N$$

    Next you ask your friend the value of the polynomial at $x = N + 1$ and say you get the answer as Q.

    Hence, we can write $Q = P(N + 1)$.

We use $M = N + 1$ and write $Q = P(M) = a_0 + a_1 M + a_2 M^2 + \cdots + a_n M^n$.

It is easy to see that if we divide Q by M, the remainder will be a_0. In the language of mathematics, we can write $Q = a_0 \bmod (M)$.

Then we determine a_1 as follows:

We can now write $Q_1 = \frac{Q - a_0}{M} = a_1 + a_2 M + a_3 M^2 + \cdots + a_n M^{n-1}$.

Now dividing Q_1 by M, the remainder will be a_1, i.e. $Q_1 = a_1 \bmod (M)$.

Proceeding this way we can determine all the coefficients. We stop when the quotient after dividing by M becomes 0.

We may note that for a monomial (polynomial with one term), it is essential to use $M = N+1$, otherwise for any polynomial having more than one terms we could have used $M = N$. But for a monomial like x^2 and x^3, $P(1)$ and $P(N)$ will be 1 in both cases and we will not be able to determine which one was written by the friend. But with $P(1)$ and $P(N + 1)$, we can determine uniquely which one was written by the friend.

Let us now demonstrate the method for the given polynomial $P(x) = 1 + 12x + 2x^3$. In this case, I know first $P(1) = N = 15$. When $Q = P(16) = 8385$.

Thus, $8385 = 16 \times 524 + 1$ and hence $a_0 = 1$.

Next we write $524 = 16 \times 32 + 12$ and hence $a_1 = 12$.

Then we write $32 = 16 \times 2 + 0$ and hence $a_2 = 0$.

In the final step (reaching zero for the quotient), we get $2 = 16 \times 0 + 2$ and hence $a_3 = 2$.

Thus, the polynomial is $P(x) = 1 + 12x + 0x^2 + 2x^3 = 1 + 12x + 2x^3$.

Alternative method: We may recall that in chapters 2 and 3 we mentioned that the decimal expression of integers and fractions are nothing but polynomials in 10 and 10^{-1}, respectively. We will use this concept in this alternative method. Towards this, first we determine the minimum value of p for which $10^p > N$, where $N = P(1)$. Then we get the second value of the polynomial at $x = (10)^{-p}$, say this value is R. Then the integer to the left of the decimal point in R is the constant term a_0 and p-digit integers represented by taking every p consecutive digits on the right of the decimal point sequentially determine the coefficients $a_1, a_2, a_3, \ldots$. Now we illustrate this method with the same example $P(x) = 1 + 12x + 2x^3$. For this polynomial $N = 15$. So the value of $p = 2$, as $10 < 15 < 100$. Next we ask the value $P(0.01)$, which we are given as $1 + 0.12 + 0.000002 = 1.120002$. Thus, we get $a_0 = 1$, $a_1 = 12$, $a_2 = 00$, $a_3 = 02$, thus the required polynomial is $P(x) = 1 + 12x + 0x^2 + 2x^3 = 1 + 12x + 2x^3$.

54. From the given values, following the methods explained in Problem number 52, we first get

$$\sum_{i=0}^{n} a_i = 4 \text{ or, } a_0 + a_1 + a_2 + \cdots + a_n = 4 \tag{i}$$

and

$$a_0 + 5a_1 + 5^2 a_2 + \cdots + 5^n a_n = 136 \tag{ii}$$

Dividing both sides of eq. (ii) by 5 we determine the remainder $a_0 = 1$.

Thus, we get $5a_1 + 5^2 a_2 + \cdots + 5^n a_n = 135$. Now dividing both sides of this equation by 5, one obtains $a_1 + 5a_2 + \cdots + 5^{n-1} a_n = 27$, if we again divide both sides of this equation by 5, the remainder determines the value of $a_1 = 2$. Following this procedure again and again we find $a_2 = 0$ and $a_3 = 1$. At this stage if we use eq. (i), it is easy to see $a_i = 0$ for all $i \geq 4$.

Hence, the polynomial is given by $P(x) = 1 + 2x + x^3$ and consequently $P(3) = 1 + 6 + 27 = 34$.

55. $2^{29} = (2^6)^4 \times 2^5 = (1 + 63)^4 \times (5 + 27)$

$= (1 + k_1 \times 63) \times (5 + 3 \times 9)$

$= (1 + 7k_1 \times 9)(5 + 3 \times 9)$

$= (5 + k_2 \times 9)$

where k_1 and k_2 are positive integers. If all the ten digits were present then the sum of the digits would have been 45 which is a multiple of 9. It is seen that if we add the digit 4 at the end of the number 2^{29}, then the number and hence the sum of the digits of that number would be divisible by 9. Thus, the missing digit is 4. This is not possible by having any other digit but 4 (nine different digits have already appeared in this nine-digit number). In the language of mathematics the solution would have been written as follows:

$$2^{29} = (2^6)^4 \times 2^5 = (1 \bmod (9))^4 \times (5 \bmod (9)) = 5 \bmod (9)$$

is known that with all the ten digits appearing once the sum of the digits would be $45 = 0 \bmod (9)$.

Hence, the missing digit is $9 - 5 = 4$. We know that if a number is divisible by 9 then the sum of its digits is divisible by 9 and vice versa.

56. First we factorize the numerator and denominator of the last term in the product as follows:

$$m^3 - 1 = (m - 1)(m^2 + m + 1)$$
$$m^3 + 1 = (m + 1)(m^2 - m + 1) = (m - 1 + 2)[(m - 1)^2 + (m - 1) + 1]$$

Then the entire product is written as

$$P_m = \frac{7}{9} \times \frac{26}{28} \times \frac{63}{65} \times \cdots \times \frac{m^3 - 1}{m^3 + 1}$$

$$= \left(\frac{1}{3} \times \frac{7}{3}\right) \times \left(\frac{2}{4} \times \frac{13}{7}\right) \times \left(\frac{3}{5} \times \frac{21}{13}\right) \times \cdots \times \left(\frac{m - 1}{m + 1} \times \frac{m^2 + m + 1}{m^2 - m + 1}\right)$$

$$= \frac{2}{3} \times \left(\frac{m^2 + m + 1}{m(m + 1)}\right) = \frac{2}{3} \times \left(1 + \frac{1}{m(m + 1)}\right)$$

Now as $m \to \infty$ and $P_\infty \to 2/3$.

57. $N = \dfrac{\left(10^4 + 324\right)\left(22^4 + 324\right)\left(34^4 + 324\right)\left(46^4 + 324\right)\left(58^4 + 324\right)}{\left(4^4 + 324\right)\left(16^4 + 324\right)\left(28^4 + 324\right)\left(40^4 + 324\right)\left(52^4 + 324\right)}.$

$$N = \prod_{n=0}^{4} \frac{(10 + 12n)^4 + 18^2}{(4 + 12n)^4 + 18^2}$$

$$= \prod_{n=0}^{4} \frac{[(10 + 12n)^2 + 18]^2 - 36(10 + 12n)^2}{[(4 + 12n)^2 + 18]^2 - 36(4 + 12n)^2}$$

$$= \prod_{n=0}^{4} \frac{[(10 + 12n)^2 + 18 + 6(10 + 12n)][(10 + 12n)^2 + 18 - 6(10 + 12n)]}{[(4 + 12n)^2 + 18 + 6(4 + 12n)][(4 + 12n)^2 + 18 - 6(4 + 12n)]}$$

$$= \prod_{n=0}^{4} \frac{(144n^2 + 312n + 178)(144n^2 + 168n + 58)}{(144n^2 + 168n + 58)(144n^2 + 24n + 10)}$$

$$= \prod_{n=0}^{4} \frac{144n^2 + 312n + 178}{144n^2 + 24n + 10}$$

Now we write $N = \dfrac{N_0 N_1 N_2 N_3 N_4}{D_0 D_1 D_2 D_3 D_4}$ and also note that

$$D_{n+1} = 144(n + 1)^2 + 24(n + 1) + 10 = 144n^2 + 312n + 178 = N_n$$

Thus,

$$N = \frac{N_4}{D_0} = \frac{144(4)^2 + 312(4) + 178}{10} = \frac{3730}{10} = 373$$

58. We can use 4 pieces weighing $3^0 (= 1)$, $3^1 (= 3)$, $3^2 (= 9)$, $3^3 (= 27)$ kilograms, respectively. Note that $1 + 3 + 9 + 27 = 40$. Moreover, it may be pointed out that using weights of integral powers of 3, we can weigh all integral kilograms up to the sum of these weights. For a single pan balance, one needs all integral powers of 2, like, 1, 2, 4, 8, 16, ... (i.e. same as the binary expression of numbers).

59. (a) We write

$$S = \sum_{n=1}^{\infty} P_n$$

where $P_n = \dfrac{2}{n(n+1)} = 2\left(\dfrac{1}{n} - \dfrac{1}{n+1}\right)$.

Hence,

$$S = 2\left(1 - \frac{1}{2} + \frac{1}{2} - \frac{1}{3} + \frac{1}{3} - \frac{1}{4} + \ldots\right) = 2 \times 1 = 2$$

(b) The given series can easily be written as

$$\frac{2^2 - 1^2}{2^2 \cdot 1^2} + \frac{3^2 - 2^2}{3^2 \cdot 2^2} + \frac{4^2 - 3^2}{4^2 3^2} + \cdots + \frac{7^2 - 6^2}{7^2 6^2}$$

$$= \frac{1}{1^2} - \frac{1}{2^2} + \frac{1}{2^2} - \frac{1}{3^2} + \frac{1}{3^2} - \frac{1}{4^2} + \cdots + \frac{1}{6^2} - \frac{1}{7^2}$$

$$= 1 - \frac{1}{7^2} = \frac{48}{49}$$

60. For the first few values of n, we write

$$n = 1, \quad S_1 = 1 = 2! - 1$$
$$n = 2, \quad S_2 = 5 = 3! - 1$$
$$n = 3, \quad S_3 = 23 = 4! - 1$$

Observing the pattern above, we guess $S_n = (n+1)! - 1$.

Now we will prove this result by the method of induction. Assuming the above guess we write

$$S_{n+1} = (n+1)! - 1 + (n+1)!(n+1)$$
$$= (n+1)! + -1 + (n+1)!(n+2-1)$$
$$= (n+2)! - 1 = (n+1+1)! - 1$$

Thus, our guess is proved to be correct. If the guess is correct for S_n, then it is correct for S_{n+1}. It is true for $n = 1$, hence for all values of n.

61. For the first few values of n, we write

$$n = 1, \quad S_1 = \frac{1}{2} = 1 - \frac{1}{2} = 1 - \frac{1}{2!} = 1 - \frac{1}{(1+1)!}$$

$$n = 2, \quad S_2 = \frac{5}{6} = 1 - \frac{1}{6} = 1 - \frac{1}{3!} = 1 - \frac{1}{(2+1)!}$$

$$n = 3, \quad S_3 = \frac{23}{24} = 1 - \frac{1}{24} = 1 - \frac{1}{4!} = 1 - \frac{1}{(3+1)!}$$

It must be mentioned that the last step in S_1 is written after looking at the next two lines giving the expressions of S_2 and S_3. Observing these patterns shown above, we guess the answer as

$$S_n = 1 - \frac{1}{(n+1)!}$$

Now as in Problem number 60, we use the method of induction assuming the above guess to be correct. Then we write

$$S_{n+1} = 1 - \frac{1}{(n+1)!} + \frac{(n+1)}{(n+2)!} = 1 - \left[\frac{1}{(n+1)!} - \frac{n+1}{(n+2)!}\right]$$

$$= 1 - \frac{1}{(n+2)!} = 1 - \frac{1}{[(n+1)+1]!}$$

So our guess turns out to be correct and

$$S_n = 1 - \frac{1}{(n+1)}$$

62. First we write

$$\frac{1}{i(k+i)} = \frac{1}{k}\left(\frac{1}{i} - \frac{1}{k+i}\right)$$

Now it is easy to write

$$S = \sum_{i=1}^{\infty} \frac{1}{i(k+i)} = \sum_{i=1}^{\infty}\left[\frac{1}{k}\left(\frac{1}{i} - \frac{1}{(k+i)}\right)\right]$$

Now we can observe that for $i > k$, every positive term cancels with an equal negative term, so the sum is

$$S = \frac{1}{k}\left(1 + \frac{1}{2} + \frac{1}{3} + \cdots + \frac{1}{k}\right)$$

63. We can write

$$\frac{1}{1.2} + \frac{1}{3.4} + \frac{1}{5.6} + \cdots + \frac{1}{(2n-1)2n}$$

$$= \left(1 - \frac{1}{2}\right) + \left(\frac{1}{3} - \frac{1}{4}\right) + \left(\frac{1}{5} - \frac{1}{6}\right) + \cdots + \left(\frac{1}{2n-1} - \frac{1}{2n}\right)$$

$$= 1 + \frac{1}{2} + \frac{1}{3} + \frac{1}{4} + \cdots + \frac{1}{2n-1} + \frac{1}{2n} - 2\left(\frac{1}{2} + \frac{1}{4} + \frac{1}{6} + \cdots + \frac{1}{2n}\right)$$

$$= 1 + \frac{1}{2} + \frac{1}{3} + \frac{1}{4} + \cdots + \frac{1}{2n-1} + \frac{1}{2n} - 1 - \frac{1}{2} - \frac{1}{3} - \frac{1}{n}$$

$$= \frac{1}{n+1} + \frac{1}{n+2} + \frac{1}{n+3} \cdots + \frac{1}{2n}$$

64. (a) First we write

$$3 = \sqrt{9} = \sqrt{1+8} = \sqrt{1+2.4} = \sqrt{1+2\sqrt{16}} = \sqrt{1+2\sqrt{1+3.5}}$$

$$= \sqrt{1+2\sqrt{1+3\sqrt{25}}} = \sqrt{1+2\sqrt{1+3\sqrt{1+4.6}}}$$

$$= \sqrt{1+2\sqrt{1+3\sqrt{1+4\sqrt{1+35}}}}$$

To show that this process can be continued for ever we write a general term starting from n (above we have started with $n = 2$)

$$\sqrt{1+n(n+2)} = \sqrt{1+n\sqrt{(n+2)^2}} = \sqrt{1+n\sqrt{n^2+4n+4}}$$

$$= \sqrt{1+n\sqrt{1+(n+1)(n+3)}}$$

We should observe that in place of the starting with $n(n + 2)$, at the end we are arriving with a term where n is replaced by $(n + 1)$ and consequently getting $(n + 1)(n + 3)$. Thus, the process can continue for ever with the value of n increasing by 1 at every step.

In 1911 Ramanujan posed this problem for the readers of a journal, but he gave only the righthand side with a? mark in the lefthand side. That makes the problem very tough, no one could answer it. After three months he gave the solution (which is not easy at all). We have converted the problem by giving the lefthand side and manipulated it to generate the righthand side.

(b) $\underbrace{999,999,\ldots,999,999}_{\text{a string of ninety three 9's}} \times \underbrace{444,444,\ldots,444,444}_{\text{a string of ninety three 4's}}$.

First we multiply and divide the above given expression by a string of ninety three 9's to get

$$\Big(\underbrace{999,999,\ldots,999,999}_{\text{a string of ninety three 9's}}\Big)^2 \times \tfrac{4}{9} = \left(10^n - 1\right)^2 \times \tfrac{4}{9},$$

where $n = 93$.

Hence, the product can be written as

$$\tfrac{4}{9} \times [(10)^{2n} - 2 \times (10)^n + 1]$$

$$= \tfrac{4}{9} \times (1\underbrace{000,000,\ldots,000,000}\ -2\underbrace{000,000,\ldots,000,000}+1)$$

a string of $2n$ number of 0's   a string of n number of 0's

$$= \tfrac{4}{9} \times (\underbrace{999,999,\ldots,999,999}8\quad 2\underbrace{000,000,\ldots,000,000}1)$$

a string of $(n-1)$ number of 9's   a string of $(n-1)$ number of 0's

$$= 4 \times (\underbrace{111,111,\ldots,111,111}10\quad \underbrace{888,888,\ldots,888,889})$$

a string of $(n-1)$ number of 1's   a string of $(n-1)$ number of 8's

$$= \quad \underbrace{444,444,\ldots,444,444}3\quad \underbrace{555,555,\ldots,555,555}6$$

a string of $(n-1)$ number of 4's   a string of $(n-1)$ number of 5's

Hence, the sum of the digits in the product is $(4+5) \times (n-1)+3+6 = 9n$.

Substituting $n = 93$, the sum of the digits in the product is $9 \times 93 = 837$.

65. Since the ten-digit number is divisible by 10, $j = 0$. Similarly, since the number $abcde$ is divisible by 5, we can conclude $e = 5$ (as the digit 0 has already been allotted to j).

Next we observe that the digits b, d, f and h must be even (out of 2, 4, 6 and 8 in some order, still to be determined) as the numbers ending with these digits are respectively, divisible by 2, 4, 6 and 8. Thus, the only remaining odd digits, viz. 1, 3, 7 and 9 must be assigned in some order (to be determined) to rest of the letters, i.e. a, c, g and h.

Thus, at this stage we know $j = 0, e = 5, (b, d, f, h) \in (2, 4, 6, 8)$ and $(a, c, g, i) \in (1, 3, 7, 9)$.

Now we consider the requirement that the number $abcdef$ is divisible by 6. Towards this end, we write $abcdef = abc \times 100 + def$. It is known that the number abc is divisible by 3, so that the part $abc \times 100$ is definitely divisible by 6.

Therefore, we conclude that the number def is divisible by 6. Since the digit f is known to be even, we infer that the number def is divisible by 3.

Hence, $(d + e + f)$ is divisible by 3, since we know that $e = 5$, we write $(d + 2 + f)$ is divisible by 3.

Next we consider that the number $abcd$ is divisible by 4, with c an odd digit. So we conclude $d \neq 4, 8$; since with any odd value of c, the number cd can then not be divisible by 4. So we consider separately the only two possibilities for d along with the condition $(d+2+f)$ divisible by 3 as noted above. Thus, the two possibilities are: (i) $d = 2, f = 8$ and (ii) $d = 6, f = 4$ which we consider below separately.

(i) If we take $d = 2$ and $f = 8$ and consider that the number *abcdefgh* is divisible by 8, i.e. the number *fgh* or *8gh* is also divisible by 8. Now $8gh = 800 + gh$, so *gh* also must be divisible by 8. We at this stage are left with two possible values for the even digit *h*, either 4 or 6($d = 2$ and $f = 8$ have already been assigned). But with the odd digits (1, 3, 7, 9) available for *g*, 4 as *h*, cannot make any possible *gh* divisible by 8. So we have to take under this circumstance $h = 6$. With this choice the possible choices for *g* reduce to either 1 or 9 and $b = 4$ (only even digit left). Thus, at this stage we can write; $e = 5, j = 0, d = 2, f = 8, h = 6, b = 4, g = 1$ or 9. We consider two cases (ia) $g = 1$ and (ib) $g = 9$ separately.

Case (ia) $g = 1$:

In this case (a, c, e) must be from the set (3, 7, 9). Now the number *abc*, i.e. *a4c* must be divisible by 3, so the sum $(a + 4 + c)$ must be divisible by 3. It can be easily checked that by taking any two values from the set (3, 7, 9) for *a* and *c*, it is not possible to satisfy the criterion for divisibility by 3. So this case (ia) is not feasible.

Case (ib) $g = 9$:

In this case to satisfy the criterion for divisibility by 3, while choosing (a, c, e) from the available set (1, 3, 7) we can have either $(a = 1, c = 7)$ or $(a = 7, c = 1)$. Accordingly, the number *abcdefg* can be either 1472589 or 7412589. But neither of these two is divisible by 7 as required.

Hence, this case (ib) is also not feasible.

So the overall conclusion is that case (i), i.e. $d = 2$ and $f = 8$ is not a solution.

Now we now proceed assuming the case (ii), i.e. $d = 6$ and $f = 4$, So the number *abcdefgh* becomes *abc654gh* and this number must be divisible by 8, so the number *4gh* is divisible by 8. Now $4gh = 400 + gh$ and hence the number *gh* must be divisible by 8. At this stage we have only two even digits, 2 and 8 to choose from for the letter *h*. But none of the four combinations, 18, 38, 78 and 98 [choosing the odd digit *g* from the available set (1, 3, 7, 9)] is divisible by 8.

Therefore, we take $h = 2$.

Hence, the last remaining even digit $b = 8$. When the ten-digit number can now be written as *a8c654g2i0*, where $(a, c, g$ and $i)$ have to be chosen from the set (1, 3, 7 and 9).

Next we observe that for $g2(= gh)$ to be divisible by 8 (as mentioned above), $g =$ either 3 or 7.

If we take $g = 3$, then for choosing $(a, c$ and $e)$ we are left with the set (1, 7 and 9). Also $(a + 8 + c)$ has to be divisible by 3. Thus, we get four possible sets for (*a* and *c*) as (1, 9), (9, 1), (7, 9) and (9, 7) from this the remaining odd digit 5 must be assigned to *e*. Accordingly, for the complete number, we get 4 possible solutions for the number *abcdefg* as 1896543, or, 9816543, or, 7896543, or, 9876543. But it can be verified that none of these four numbers is divisible by 7 as required.

Hence, we conclude that *g* = 7, when for choosing (*a*, *c* and *e*) we are left with the set (1, 3 and 9). For (*a* + 8 + *c*) to divisible by 3 the four possible set for (*a* and *c*) turn out as (1, 3), (3, 1), (1, 9), (9, 1). Now considering four possible values of the number *abcdefg* come out as 1836547, or, 3816547, or, 1896543, or 9816547. Out of these only the number 3816547 is divisible by 7.

Hence, the unique answer to the problem is determined as 3816547290.

66. First we observe $2^y > 615$, so $y > 9$ as $2^9 (= 512) < 615 < 2^{10} (= 1024)$.

Now we consider the possible values of the digit in the unit's place on both sides of the given equation (these must be the same). First we note that for all possible ten digits in the unit's place of *x* and the corresponding digits in x^2 and $x^2 + 615$. We also note that in integral powers of 2 the unit's digit changes periodically (for $y > 0$) in the cycle 2, 4, 8, 6, 2, 4, ...

Unit's digit in

x	0	1	2	3	4	5	6	7	8	9
x^2	0	1	4	9	6	5	6	9	4	1
$x^2 + 615$	5	6	9	4	1	0	1	4	9	6

Now equating the unit's digit of both sides of the given equation, we conclude that this digit has to be either 4 or 6. This in turn implies that *y* has to be even, and further that the unit's digit of *x* must be one from the set (1, 3, 7, 9).

Let us write $y = 2m$, with *m* as an integer since *y* is known to be even. At this stage let us write the given equation as

$$615 + x^2 = 2^y = 2^{2m} = (2^m)^2$$
$$\text{or, } 615 = (2m)^2 - x^2 = (2^m + x)(2^m - x)$$

Now we note that there are three prime factors of 615, viz. 3, 5 and 41.

Hence, 615 can be factored (with 2 factors, writing the bigger factor first) in four different ways as 615×1, 205×3, 123×5 and 41×15. We list below the values of *x* and *n* by equating the factors $(2^m + x)$ with the bigger number and $(2^m - x)$ with the smaller numbers:

$2^m + x$	$2^m - x$	2^m	m	x
615	1	308	?	307
205	3	104	?	101
123	5	64	6	59
41	15	28	?	13

It is seen that only integral value for m is obtained as 6, when $y = 2m = 12$ with $x = 59$. It must be observed that $y > 9$ as required and also the digit in the unit's place of x is 9 (i.e. one of the values from the set (1, 3, 7, 9) as mentioned earlier.

67. Since $a_{i+1} > a_i$ for all values of i, we get

$$6 \sum_{i=7}^{10} a_i > 24a_7 \text{ and also } 4 \sum_{i=1}^{6} a_i < 24a_7$$

Thus,

$$6 \sum_{i=7}^{10} a_i > 4 \sum_{i=1}^{6} a_i$$

Now we add $6 \sum_{i=1}^{6} a_i$ to both sides of this inequality to get

$$6 \sum_{i=1}^{10} a_i > 10 \sum_{i=1}^{6} a_i \text{ or, } \frac{\sum_{i=1}^{6} a_i}{6} < \frac{\sum_{i=1}^{10} a_i}{10}$$

68. (a) First we substitute $y_n = \tan x_n$; $n = 1, 2, 3, \ldots, 7$. For $-\infty < y < \infty$, $-\pi/2 < x < \pi/2$.

If we consider seven values of the variable x, within the range π, from $-\pi/2$ to $\pi/2$, then there must be two values of the variable x, say, x_i and x_j, which differ by less than $\pi/6$. This follows from the pigeonhole principle, mentioned earlier. Think of six intervals (holes) each measuring $\pi/6$ to cover the entire range $-\pi/2$ to $\pi/2$. Now if we put seven values of the variable in these holes, at least two of them must be in the same hole. So we can write $0 \le (x_i - x_j) \le (\pi/6)$.

Also note that the value of $\tan x$ increases monotonically as x increases from $-\pi/2$ to $\pi/2$. Hence, we get

$$\tan 0 \le \tan(x_i - x_j) \le \tan(\pi/6) \text{ or, } 0 \le \frac{\tan x_i - \tan x_j}{1 + \tan x_i \tan x_j} \le \frac{1}{\sqrt{3}}$$

$$\text{or, } 0 \le \frac{y_i - y_j}{1 + y_i y_j} \le \frac{1}{\sqrt{3}}$$

(b) Again we apply the powerful pigeonhole principle mentioned in part (a) to solve this problem. With 10 elements, the total number of subsets is $2^{10} = 1024$. Discounting the null set and the universal set the total number of subsets is 1022. The lowest possible sum of a subset is 1, and the highest possible sum of remaining number of 9 elements is $(92 + 93 + 94 + \cdots + 100) = 864$. So there are only 864 different sums (pigeonholes), but the number of possible subsets (pigeons) is 1022.

Hence, there must be more than 1 subsets that will have the same sum. So there must be at least two subsets with the same sum. If there are some common elements in these two subsets, remove them from both the sets to have two disjointed subsets having the same sum.

69. If $x = 1$, then $\dfrac{x^{2013} + 1}{1 + x + x^2} = \dfrac{2}{3}$, so the remainder is 2. If $x \neq 1$, then

$$x^3 = 1 \Leftrightarrow (1 + x + x^2) = 0$$

Now we can write

$$x^{2013} + 1 = q(x)(1 + x + x^2) + r(x)$$

Our objective is to find this remainder $r(x)$. Towards that end, we substitute $x^3 = 1$ on the lefthand side and $(1 + x + x^2) = 0$ on the right.

$$r(x) = (x^3)^{671} + 1 = 1^{617} + 1 = 1 + 1 = 2$$

Hence, the remainder is always 2.

70. $\displaystyle\sum_{k=1}^{n} \frac{kx^k}{1 + x^{2k}} = \frac{n(n + 1)}{4} = \frac{1}{2}\frac{n(n + 1)}{2} = \frac{1}{2}\sum_{k=1}^{n} k$

or, $\displaystyle\sum_{k=1}^{n} k\left(\frac{x^k}{1 + x^{2k}} - \frac{1}{2}\right) = 0$ or, $\displaystyle\sum k\frac{(2x^k - 1 - x^{2k})}{2(1 + x^{2k})} = 0$

or, $-\displaystyle\sum_{k=1}^{n} \frac{k(x^k - 1)^2}{2(1 + x^{2k})} = 0$

Hence, for all value of k

$$x^k - 1 = 0 \rightarrow x = 1$$

71. We know that (i) $a^0 = 1$, if $a \neq 0$; (ii) $1^a = 1$; (iii) $(-1)^{\text{any even integer}} = 1$. Using this information in the given equation (with the righthand side equal to 1) separately we write

(i) $x^2 - 13x + 42 = 0$ or, $(x - 7)(x - 6) = 0$, or, $x = 7, x = 6$

(ii) $x^2 - 7x + 11 = 1$ or, $x^2 - 7x + 10 = 0$ or, $(x - 2)(x - 5) = 0$ or, $x = 2, x = 5$

(iii) $x^2 - 7x + 11 = -1$ or, $x^2 - 7x + 12 = 0$ or, $(x - 3)(x - 4) = 0$ or, $x = 3$, $x = 4$.

Now we verify that, for these last two values of $x(= 3$ and $4)$, the value $x^2 - 13x + 42$ comes out as 12 and 6, respectively, and both are even as required.

So the solutions are $x = 2, 3, 4, 5, 6$ and 7.

72. We know that for a set of numbers, x_i, $i = 1, 2, 3, \ldots, n$, the arithmetic mean is greater than equal to the geometric mean or,

$$\frac{\sum_{i=1}^{n} x_i}{n} \geq (x_1 x_2 x_3 \ldots x_n)^{1/n}$$

The equality sign holds when the values of all the numbers are same, i.e. $x_1 = x_2 = x_3 = \cdots = x_n$. In other words, for a prescribed value of the sum the product will be maximum when

$$x_1 = x_2 = x_3 = \cdots = x_n = \frac{\sum_{i=1}^{n} x_i}{n}$$

So in this problem we have to determine the value of n when $(271/n)^n$ reaches the maximum value. To determine this value of n, first note that the value of $(271/n)^n$ initially increases with increasing n (check with $n = 1$ and 2). So we determine this optimum value of n by ensuring

$$(271/n)^n > (271/(n + 1))^{n+1} \text{ or, } n\left(1 + \frac{1}{n}\right)^{n+1} > 271$$

For a sufficiently large value of n, we know that

$$\left(1 + \frac{1}{n}\right)^{n+1} \approx e$$

Thus, we get $n > \frac{271}{e}$; substituting the approximate value of

$$e \approx 2.718281828\ldots$$

we get n is approximately greater than $99.7\ldots$.

Now we try $n = 99$ and 100 to observe that

$$(271/100)^{100} > (271/99)^{99}$$

Therefore, we divide 271 by 100 to get 100 terms each equal to 2.71 when the product of the numbers will be maximum with the sum 271.

73. The problem essentially states that the number of students is the smallest number that can be divided by 12 different numbers defining the number of rows. Thus, we have to determine the smallest number which has exactly 12 divisors.

Now $12 = 12 \times 1 = 6 \times 2 = 4 \times 3 = 3 \times 2 \times 2$.

Hence, by considering the prime factorization, the smallest numbers in each category with exactly 12 divisors are: (i) 2^{11}, (ii) $2^5 \times 3$, (iii) $2^3 \times 3^2$ and (iv) $2^2 \times 3 \times 5$.

Out of these four, obviously the smallest number is the last one, which is 60. The 12 divisors of 60 are 1, 2, 3, 4, 5, 6, 10, 12, 15, 20, 30 and 60.

74. If A and B belong to the same row then obviously B is taller. If the belong to different rows, then let us imagine that A belongs to the mth row and nth column and B to the pth row and qth column. Let us call the student standing in the pth row and nth column as C. Then it is obvious that B is taller than C and C is taller than A. Then B is certainly taller than A. Thus, in all situations B is taller than A.

75. LCM of 3, 4 and 5 is 60. So on the day number 60 all three granddaughters simultaneously call for the first time in 2017 (just like December 31, 2016). So we consider a cycle of 60 days. In this interval, one of them calls 20(= 60/3) times, the second one 15(= 60/4) times and the third one 12(= 60/5) times, the grandmother received in total 20 + 15 + 12 = 47 calls. But the number of days she received calls is less than 47, since there were days when more than one granddaughter called. Now we determine the number of such days as follows:

   (i) There are 4 days (every 12th day) when the first two granddaughters called.

   (ii) There are 3 days (every 15th day) when the first and the third granddaughter called.

   (iii) There are 2 days (every 20th days) when the second and the third granddaughter called.

   (iv) On the 60th day all three called.

Thus, the number of days when the grandmother received calls is 47 − 4 − 3 − 2 − 2 = 36. So there are 60 − 36 = 24 days she did not receive any call.

Hence, in 360 days she did not receive any call on 24 × 6 = 144 days.

In the remaining 5 days she did not receive any call on the first 2 days and then she received on the remaining three days. Thus, in total, she did not receive any call on 144 + 2 = 146 days.

Alternative method: Again we consider a period of 60 days. During this period, she did not receive any call from the first granddaughter on 60 ×

(2/3) = 40 days. In these 40 days she did not receive any call from the second granddaughter on 40 × (3/4) = 30 days. In these 30 days she did not receive any call from the third granddaughter on 30 × (4/5) = 24 days.

Hence, on 24 days during this 60-days interval, she did not get any call on 24 days as determined in the first method. We could have directly considered 360-days period to get 144 days of no call arriving.

76. It is obvious that the cells visited even number of times will remain closed and those visited odd number of times will remain open. It may be noted that factors of a number occur in pair as $a \times b = b \times a$, except for perfect square numbers. Since $m^2 = m \times m$, the factor m does not have a pairing value.

Hence, the cell numbers which are perfect squares, like 1, 4, 9, 16, 25, 36, 49, 64, 81 and 100 will remain open after all the visits are over.

77. In Fig. 5.1, the numbers 1, 2, 3, ... indicate the positions of the girls announcing these numbers. Let the girl announcing the number 6 gave the number x to her two neighbours indicated by the numbers 5 and 7. It is easy to see by considering the girl announcing 5, that the girl announcing the number 4 had given the number $(10 - x)$ to her two neighbours. Proceeding counterclockwise along the circle in this fashion, as indicated in the figure, we finally get that the girl announcing the number 7 received the number $(x + 12)$ from the girl who announced the number 8. Now we can write $(x + x + 12)/2 = 7$ or, $x = 1$.

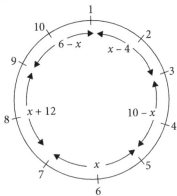

Fig. 5.1: Solution No. 77

78. Let L m be the length of the circular track and B ran x m when they met for the first time. Then it is easy to see that $300 + x = L/2$.

They both ran double the distance when they met for the second time as they together covered now the full length of the track. Thus, $600 + 2x = L$.

It is given that, $2x = 400$ m. Hence, $L = 600 + 400 = 1{,}000$ m.

79. For the three people to meet the distance covered by them should differ by integral multiple of the track length, Moreover, the distance covered by them at any time will be in the ration 9 : 8 : 7. So the first time they meet after A covers 9 rounds (as 9, 8 and 7 are co-primes). So they meet after $9 \times 5 = 45$ minutes.

    Alternative method: When C has walked 700 m, A is 200 m ahead of C; B is 100 m ahead of C. So for B to be ahead by 1 km from C, the distance covered by C has to be 7 km, i.e. C has to make 7 rounds. In the same time A covers 2 km more than C (and also 1 km more than B), i.e. A covers 9 rounds.

80. If there are 2 people then the person numbered 2 goes away and the person numbered 1 remains. If there are $4(= 2^2)$ people to start with, then people numbered 2 and 4 go in the first round. So only two people, numbered originally 1 and 3 remain. In the next round the person numbered 3 goes away.

    Hence, again the person originally numbered 1 remains. It is easy to realize that if we start with 2^n people, for all (positive integer) values of n, finally the person originally numbered 1 remains. Thus, if there are S number of people to start with where $2^n < S < 2^{n+1}$, then $S = 2^n + R$; when $R(< 2^n)$ number of people leave, we are left with 2^n number of people. The last person leaving was occupying originally $(2R)$th position. So renumbering the $(2R + 1)$th person as 1, he will be the last person remaining.

    For $S = 1000$, we write

    $$2^9 (= 512) < S < 2^{10} (= 1024).$$

    Hence, $R = 1000 - 512 = 488$. So the last person to remain was originally numbered $2R + 1 = 2 \times 488 + 1 = 977$.

81. Let us assume that with initially n people standing in the circle, the person standing originally in the pth position remains until the end. Now if we add an extra people to the initial list, i.e. start with $(n + 1)$ persons then the last remaining person would have originally occupied $(p + 3)$th position, if $(p + 3) \leq (n + 1)$ and if $(p + 3) > (n + 1)$, then the originally occupied position of the last remaining person would have been $[(p + 3) - (n + 1)]$. If we proceed in this way, then we conclude that had we started with $(n + x)$ people then the originally occupied position of the last remaining person is given by $y = (p + 3x) - (n + x) = (p + 2x - n)$. Since this equation is to be used repeatedly, we write it separately below:

    $$y = p + 2x - n \tag{i}$$

Now we determine some special values of n for which the value of y comes out as either 1 or 2. Here onward the values of n for the first category (with $y = 1$) will be denote by n_1 and those in the second category (with $y = 2$) will be denoted by n_2. It can be easily verified that the minimum value of $n_1 = 4$ (when successively, people originally standing at positions 3, 2 and 4 leave), thus $p = 1$. Similarly, one gets the minimum value of $n_2 = 2$ (when counting 1, 2, 1, the first person leaves), with $p = 2$.

Now starting with these values of n's and p's and using eq. (i) we find the minimum values of x to yield $y =$ either 1 or 2. In the next step, we get the new value of n by increasing the current value by x and the new value of p will be given by the value of y. The details of the first few steps are worked out below:

First we take

$$n_1 = 4, \ p = 1 \rightarrow x = 2, \ y = 1$$

so in the next step

$$p = 1, \ n_1 = 4 + 2 = 6 \tag{ii}$$

Now

$$n_1 = 6, \ p = 1 \rightarrow x = 3, \ y = 1$$

so in the next step

$$p = 1, \ n_1 = 6 + 3 = 9 \tag{iii}$$

Then

$$n_1 = 9, \ p = 1 \rightarrow x = 5, \ y = 2$$

so in the next step

$$p = 2, \ n_2 = 9 + 5 = 14 \tag{iv}$$

Then we start with the minimum value $n_2 = 2$ and proceed as follows:

$$n_2 = 2, \ p = 2 \rightarrow x = 1, \ y = 2$$

so in the next step

$$p = 2, \ n_2 = 2 + 1 = 3 \tag{v}$$

Then

$$n_2 = 3, \ p = 2 \rightarrow x = 1, \ y = 1$$

so in the next step

$$p = 1, \ n_1 = 3 + 1 = 4 \tag{vi}$$

We will now show just one more step, following (vi), when we have $n_2 = 14, \ p = 2 \rightarrow x = 7, \ y = 2$, so in the next step, $p = 2, \ n_2 = 14 + 7 = 21$.

Proceeding in this manner recursively, we make the Table shown below.

	$y = 1$	$y = 2$
n_1 and n_2	4	2
	6	3
	9	14
	31	21
	70	47
	105	158
	355	237
	799	533
	1798	1199
	2697	4046
	9103	6069

This table can be used to solve the problem up to having 10,000 people to start with. For example, we have initially 1,000 people. Observing this table we note that the number closest to but less than 1,000 is 799. So after removing (1000 − 799) = 201 people, we will be left with 799 people. When 201 people are removed the last person to move out was occupying 3 × 201 = 603rd position. So if we renumber the next person as 1, and as we are left with exactly 799 people, the above table suggests that the person (now) occupying the first position remains till the end. Thus, the person originally occupying 604th place remains until the end. If suppose initially there were 1,200 people, then we should consider removing just 1 person to reach the number 1199 and in that case we need to renumber the 4th person as number 1 and the person now numbered 2, i.e. originally the person numbered 5 remains until the end.

82. 10 different digits can be arranged 10! ways in 10 places. However, in this case four 4's, three 3's and two 2's can be interchanged without changing the number. So the number of different numbers is $\frac{10!}{4!\,3!\,2!}$ = 12,600.

83. Since all 4 Bengali books will be together, let us put these in a box and we move the box so that the Bengali books remain together. So we have 6 different things to arrange, 1 box, 3 English and 2 Hindi books. These 6 things can be arranged 6! ways. Now in each of these arrangements, 4 Bengali books within the box (now imagine the box is open) can be arranged in 4! different ways (but these books still stay together). So the total number of arrangements is 6! × 4! = 720 × 24 = 17,280.

84. Because 2 O's must remain together, we put these two in a box, and treat the box as one letter. Thus, there are 5 letters, viz. 1 box, 2 B's, 1 A and 1 N. So they can be arranged $\frac{5!}{2!}$ = 60 different ways, since 2 B's can be interchanged without any distinction. Now in some of these 60 arrangements 2

B's are adjacent and we have to subtract the number of such arrangements from 60 to fulfil all the conditions. To find in how many of these 2 B's are adjacent, we put 2 B's also in another box. So now there are 4 letters, viz. 1 box containing 2 O's, another containing 2 B's, 1 A and 1 N. So these can be arranged in 4!(= 24) different ways. So the total number of arrangements fulfilling all the conditions is 60 − 24 = 36.

85. Note that there are 9 single digit natural numbers, 90 two-digit numbers and 900 three-digit numbers. The number of pages in the book must be a four-digit number, because to print up to 999, one needs $9 + 2 \times 90 + 3 \times 900 = 2,889$ digits. Also to print a book with five-digit page numbers one requires many more digits than 3,189. Let the number of pages in the book be $999 + p$. Then the required number of digits to print all the page numbers is $2889 + 4 \times p = 3189$ or, $p = 75$.

Hence, the book has $999 + 75 = 1074$ pages.

86. Let the ith volume have n_i number of pages. Then we can write

$$n_2 = n_1 + 50 \tag{i}$$

$$n_3 = 1.5n_2 \tag{ii}$$

$$1 + (n_1 + 1) + (n_1 + n_2 + 1) = 1079 \text{ or, } 2n_1 + n_2 = 1076 \tag{iii}$$

Solving these three equations, we get $n_1 = 552$, $n_2 = 602$ and $n_3 = 903$.

Hence, the last page number of the book is $552 + 602 + 903 = 2057$. The prime factorization of this number is $2057 = 11 \times 11 \times 17$, so the largest prime factor is 17.

87. Let the last page number be n and the page numbers on the torn leaf be x and $(x + 1)$. As per the normal practice, we assume that the book starts with the first page on the righthand side. In other words, we look for x as an odd number. We can write

$$\sum_{k=1}^{n} k - x - (x + 1) = 15,000$$

$$\text{or, } \frac{n(n + 1)}{2} - 2x - 1 = 15,000 \tag{i}$$

It is easy to accept that $n^2 \gg n$ and x, so first we obtain an approximate value of n from (i) neglecting both n and x in comparison to n^2. Thus, we get $n^2 \approx 30,000$ or, $n \approx 173$.

Substituting $n = 173$ in (i), we get $x = 25$, i.e. an odd number as desired. In that case the page numbers on the torn leaf are 25 and 26. However, 173 was an approximate value of n, so we should try with $n = 172$ and 174. With $n = 172$, the value of x turns out to be negative, so there is no need to try with any number less than or equal to 172. Now with $n = 174$, x comes out as 112, which is even and hence rejected. With $n \geq 175$, the value of x comes out $> n$, hence these are also rejected. So the page numbers on the torn leaf are 25 and 26.

88. (a) It should be noted that initially the number of white balls is even. The number of white ball either decreases by 2 (when both are white) or does not decrease at all, as when the two balls drawn out are of opposite colours, i.e. 1 white ball comes out, 1 white ball is put in. Thus, the number of white balls remaining in the bag always stays even. So this number can never reach 1, that implies the last remaining ball cannot be white.

(b) You can reach the nth stair from either $(n - 1)$th or $(n - 2)$th stair. So if we denote the number of ways of reaching the nth stair as $F(n)$, then obviously $F(n) = F(n - 1) + F(n - 2)$, with $F(1) = 1$ and $F(2) = 2$. Recall the Fibonacci-Hemachandra sequence.

Hence, $F(12)$ will be the 13th Fibonacci number or the 12th Hemachandra number 233.

89. The sum of the numbers originally written on the paper is

$$S = \frac{2018(2018 + 1)}{2} = 1009 \times 2019$$

which is an odd number. At every step the sum decreases by twice the smaller number (between a and b) as the sum first decreases by $(a + b)$ and then increases by $|(a - b)|$. So the sum of the remaining number(s) can never be even so the last number cannot be zero.

90. We denote five people by A, B, C, D and E who, respectively takes 1, 3, 6, 8 and 13 minutes to cross the bridge. To minimize the time, we try to use the faster two of these people, i.e. A and B as many times as we can and always the fastest of the people to return alone with the torch from the destination side of the bridge. Accordingly we make the following table:

Description of the movement	Time taken in mins	Person(s) on the destination side
A and B cross the bridge	3	
A returns with the torch	1	B
D and E cross the bridge	13	
B returns with the torch	3	D and E
A and C cross the bridge	6	
A returns with the torch	1	C, D and E
A and B cross the bridge	3	A, B, C, D and E

So the total time taken is $3 + 1 + 13 + 3 + 6 + 1 + 3 = 30$ mins.

91. In Fig. 5.2, S and H indicate, respectively, the station and the home. M indicates the location (in between S and H), where the person meets his wife with the car. The dashed U shaped line shows the distance which the car did not have to travel because of the meeting in between S and H. This saving in travel by the car has saved 10 minutes of time in the time of arrival at H. So half of this travel corresponds to 5 minutes. The car was supposed to reach the station at 4 pm so the time of meeting is 3.55 pm. The man started walking at 3 pm, hence he walked for 55 minutes.

Fig. 5.2: Solution No. 91

92. Let us assume in some undefined units the distance between A and B is $5 \times 7 = 35$ units. So the speeds along downstream and upstream are, respectively 7 and 5 units/hr. So their average is the speed of the steamer is 6 units/hr and the speed of the current is 1 unit/hr hence the wooden plank will reach from A to B in $35/1 = 35$ hours.

93. Fig. 5.3 shows that the boat starting from the southern bank meets the other boat first time at the location indicated by A. During the return journey the two boats meet at the location indicated by B. Note that the 10 minutes the boats rested at the banks is of no consequence. By following the paths indicated by the dashed line it is readily seen that between the start and the first meeting the total distance covered by both the boats together is equal to width of the river. Again the total distance travelled by both the boats together between the start and the second crossing is three times the width of the river. Between the start and the first meeting the first boat travelled 400 m, so the total distance travelled by this boat between the start and second meeting is 3×400 m = 1,200 m. Since the location of B is 200 m away from the northern bank, the width of the river is easily seen to be $1,200 - 200 = 1,000$ m.

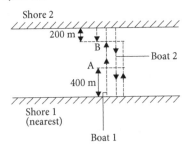

Fig. 5.3: Solution No. 93

94. We find the time taken to reach the destination by different modes of travel as

(i) by walking $60 \times (9/5) = 108$ minutes

(ii) by cycle $60 \times (9/9) = 60$ minutes

(iii) by motor cycle $60 \times (9/15) = 36$ minutes.

So the average time taken to complete 3 journeys by using three modes is $(108 + 60 + 36)/3 = 204/3 = 68$ minutes. So the journey with all three vehicles and the persons starting and finishing simultaneously cannot be accomplished in less than 68 minutes. Now we discuss how to distribute the vehicles to attain this optimum journey in 68 minutes. Let the first friend travel initial x km by motor cycle and leave the vehicle for one of his friends to pick it up and walk the rest $(9 - x)$ km. If he takes 68 minutes to complete his journey, then

$$\frac{60}{15}x + \frac{60}{5}(9 - x) = 68 \text{ or, } 4x + 108 - 12x = 68 \text{ or, } x = 5 \text{ km}$$

The second friend travels initial y km on cycle and leaves it there for the third friend to pick it up. Then he walks $(5 - y)$ km to pick up the motor cycle and rides it up to the destination. If this journey also takes 68 minutes, then we can write

$$\frac{60}{9}y + \frac{60}{5}(5 - y) + \frac{60}{15}(9 - 5) = 68 \text{ or, } \frac{20}{3}y + 12(5 - y) + 4 \times 4 = 68$$

$$\text{or, } y = \frac{3}{2} \text{ km}$$

The third friend has walked $(3/2)$ km and picked up the cycle and rides it up to the destination. To complete his journey he takes

$$\frac{60}{5}\left(\frac{3}{2}\right) + \frac{60}{9}\left(9 - \frac{3}{2}\right) = 18 + 50 = 68 \text{ minutes}$$

Thus, everybody completes the journey in 68 minutes and both the vehicles also reach the destination.

95. Fig. 5.4 shows three different instants of times in parts indicated by (a), (b) and (c). In these figures A_1, A_2 and A_3 denote the locations of the person walking in the direction of the train. Similarly, B_1, B_2 and B_3 denote those of the person walking in the opposite direction of the train. F and E denote, respectively, the front and rear ends of the train. In part (a), both persons and F are at the same location O_1 on the platform. In part (b), both persons have moved by 20 m from O_1, when E coincides with B_2 and this location is denoted by O_3 on the platform. The location of A_2 is denoted by the point O_2 on the platform. In part (c), E coincides with A_3 and this location is indicated by the point O_4 on the platform. It is given that the distance $O_1O_4 = A_1A_3 = 30$ m. So the distance $O_2O_4 = 30 - 20 = 10$ m. So between the instants shown in parts (b) and (c), A has moved 10 m and the train (note the movement of the point E between these two parts) has moved $20 + 30 = 50$ m. Thus, the train moves 5 times faster than the persons. So during the instants shown in parts (a) and (c), the train has moved $30 \times 5 = 150$ m. So F has moved 150 m from O_1 during this time interval. So in part (c) the distance between F and E is $150 - 30 = 120$ m.

Hence, the length of the train is 120 m.

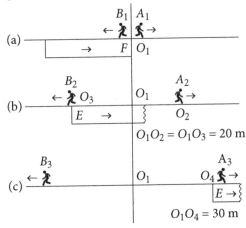

Fig. 5.4: Solution No. 95

96. Let the number of stairs that can be seen when the escalator is stationary be n. We assume that during the downward movement of the person, the number of stairs that vanishes (from the bottom, also reappears from the top) is x. During the upward movement, the speed of stepping was 5 times faster which implies the person should have made 250 steps during the time taken for the downward journey. But he made only 125 steps, which means he took half the time while going up as compared to the downward journey.

   Hence, the number of stairs that vanished during the upward journey is $(x/2)$.

   Therefore, we can write $n = 50 + x = 125 - (x/2)$, or, $x = 50$, so, $n = 50 + x = 100$.

97. Let the person go out at x minutes past 4 pm and return at y minutes past 5 pm. We also know that the minutes hand rotates 12 times faster than the hour hand. So we can write the following two equations

$$y - 20 = \frac{x}{12} \text{ and } x - 25 = \frac{y}{12}$$

Solving these two equations, we get

$$x = \frac{3840}{143}, \quad y = \frac{3180}{143}$$

So the person was out for $60 - (x - y) = 55\frac{5}{13}$ minutes.

98. For solving this problem, first we draw the graphs of distance covered versus time for the four vehicles. As the vehicles move with different but constant speeds, these graphs are represented by straight lines as shown in Fig. 5.5. We take the origin at the time (8 am) and location of the crossing of the cars A and B. Note that the slopes of the graphs for cars A, B and C are all positive and that of D is negative. The time and location of the crossings of the cars A, B and C with those of the car D are denoted by the points a, b and c, respectively. The times corresponding to these three points are 10 am, 12 noon and 2 pm, respectively. We have to find the time of the crossing of the cars B and C, i.e. the time indicated by the point G in this figure. By noting the times corresponding to the points, a, b and c it is readily seen that $ab = bc$ and $oe = ea$.

Hence, the lines ob and ce are the two medians of the triangle oac.

Hence, the point G is the centroid of the triangle oac. Thus, $oG = (2/3)ob$, when we can say $oT_G = (2/3)(12 - 8)$ hours = 2 hours 40 minutes.

Therefore, the time of crossing of the cars B and C is given by 10.40 am.

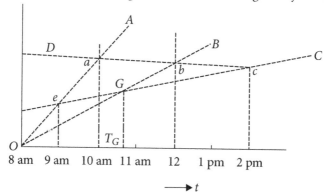

Fig. 5.5: Solution No. 98

99. Dividing 72 by 2, we get 36. Hence, Ram was born in $1972 - 36 = 1936$. Again $172/2 = 86$, his father was born in $1972 - 86 = 1886$.

100. The perfect square numbers near but less than 1950 are $1936 = 44^2$ and $1849 = 43^2$. If the person is 36 years old in 1936, then he is 50 years old in 1950. On the other hand, if he is is 43 years old in 1849, then his age in 1950 comes out as $1950 - 1806 = 144$ years, which is not possible. So he is 50 years old in 1950.

101. Let the current ages of Ram and his father be x and y years, respectively. So we can write u

$$100x + y = u^2 \qquad \text{(i)}$$
$$\text{and } 100(x - 25) + y - 25 = v^2 \qquad \text{(ii)}$$

Subtracting (ii) from (i), we get

$$u^2 - v^2 = 2525$$
$$\text{or, } (u - v)(u + v) = 5 \times 505 = 25 \times 101 = 1 \times 2525$$

Since u^2 and v^2 both are four-digit numbers, $u < 100$ and $v < 100$ and hence $u + v < 200$. Observing the above products we may conclude $u + v = 101$ and $u - v = 25$.

Thus, $u = 63$, when $u^2 = 3969$. So the current ages of Ram and his father are 39 and 69 years, respectively. You may verify 25 years ago their ages were 14 and 44, respectively, and $1444 = 38^2$ is a perfect square.

102. First we list all the different ways in which the product of three natural numbers can result in 36. In each case we also write the sum of the three numbers.

Expression of 36 as the product of three natural numbers	Sum of these three numbers
$1 \times 1 \times 36$	$1 + 1 + 36 = 38$
$1 \times 2 \times 18$	$1 + 2 + 18 = 21$
$1 \times 3 \times 12$	$1 + 3 + 12 = 16$
$1 \times 4 \times 9$	$1 + 4 + 9 = 14$
$1 \times 6 \times 6$	$1 + 6 + 6 = 13$
$2 \times 2 \times 9$	$2 + 2 + 9 = 13$
$2 \times 3 \times 6$	$2 + 3 + 6 = 11$
$3 \times 3 \times 4$	$3 + 3 + 4 = 10$

It can be easily observed that except in two cases, the sum of the three numbers is different. So even after knowing the sum, i.e. the house number, the inability of the census officer to determine the individual ages implies the sum must have been 13. In that case, two possibilities exist, (6, 6, 1) and (9, 2, 2). Since there is one eldest child, the ages are 9, 2 and 2. Otherwise there is no eldest child; there is only one youngest child and two elder (twin) children.

103. If we can dispense three consecutive numbers of chocolates using two types of packets, then there is no need to check any further. We can always add the smaller packet to these three numbers to cover the next three numbers and the process can continue forever. Now we may note that we cannot dispense 14 chocolates using the two packets in any combination. Next we check 14 + 3 = 17 chocolates also cannot be dispensed. But 17 + 3 = 20 chocolates can be sold; then we check the previous two numbers 19 and 18 also can be sold as shown now. We note that 20 = 2 × 10, 19 = 10 + 3 × 3, 18 = 6 × 3. So we need not check any further. 17 is the highest number of chocolates that cannot be purchased.

In general, for two co-prime (not having a common factor) numbers (m, n), the highest number that cannot be expressed as any linear combination of m and n is known as *Frobenius number*, given by $mn - m - n$. In the above example 3 and 10 are co-prime numbers, hence the highest number of chocolates that cannot be purchased is $3 \times 10 - 3 - 10 = 17$ as obtained earlier. However, there is no such method available for three given numbers.

104. Here if six consecutive numbers of chocolates can be dispensed as some combination of the given three packets, then we need not check any further. In this problem we observe, that 7 chocolates cannot be sold. Then we check 7 + 6 = 13, 13 + 6 = 19, 19 + 6 = 25, 25 + 6 = 31, 31 + 6 = 37 and 37 + 6 = 43, all these numbers of chocolates cannot be sold. But 43 + 6 = 49 chocolates can be sold. Then we check that all five preceding numbers can also be dispensed as shown now: 49 = 2 × 10 + 9, 48 = 8 × 6, 47 = 20 + 3 × 9, 46 = 2 × 10 + 6, 45 = 5 × 9, 44 = 1 × 20 + 4 × 6.

Hence, we conclude 43 is the highest number of chocolates that cannot be purchased.

105. Let the number of chocolates taken by different ladies be as mentioned below:

Mrs Acharjee $-A$ Mrs Banerjee $-B$
Mrs Chatterjee $-C$ Mrs Mukherjee $-M$.

So we can write the following two equations

$$A + B + C + M = 14 \tag{i}$$
$$\text{and } A + 2B + 3C + 4M = 44 - 14 = 30 \tag{ii}$$

Subtracting (i) from (ii), we get $B + 2C + 3M = 16$ or, $B + 3M = 16 - 2C$. Now we can see that B and M, must be both odd or both even. It can be verified that all the numbers lie between 2 and 5. Satisfying all the conditions, we get $M = 2$, $B = 4$, $C = 3$ and $A = 5$.

Hence, the surnames are as follows:

Mrs Sita Acharjee   Mrs Neeta Banerjee
Mrs Mita Chatterjee   Mrs Geeta Mukherjee.

106. Let the cost of 72 oranges be ₹ $x67.9y$. If this amount is expressed in paise, we get $x679y$. This number is divisible by 72 and 72 = 8 × 9, hence this

number is divisible by 8. In other words we can say the number $79y$ is divisible by 8. Thus, we conclude $y = 2$, since 792 is the only number in the range 790 to 799 that is divisible by 8. Now we can say that the number $x6792$ is divisible by 72 and hence by 9. Applying the divisibility condition for 9, we get $x + 6 + 7 + 9 + 2 = x + 24$ is divisible by 9. Thus, $x = 3$. So the total price for 72 oranges is ₹ 367.92, when the price of each orange is $367.92/72 = ₹ 5.11$.

107. Let the units digit of the total number of cows N be b.

Hence, we can write $N = 10a + b$, where a is some unknown number. Since each cow costs ₹ N, the total money received is

$$N^2 = (10a + b)^2 = 100a^2 + 20ab + b^2$$
$$= 20a(5a + b) + b^2$$

Note that the number $20(5a + b)$ is divisible by 20. So when this amount of money has been divided between the two brothers taking ₹ 10 at a time, and only b^2 rupees are remaining, the brothers had taken equal amount of money. Now according to the problem, the amount b^2 can be either 16 or 36 as only odd number of tens could be taken from this amount leaving a remainder less than 10. Note that all other perfect squares less than 100 are 1, 4, 9, 25, 49, 64 and 81. So we conclude at the end only ₹ 6 was left. So the price of the pen is ₹ $(10 - 6)/2 = ₹ 2$.

108. Suppose the cheque was written for Rupees x and Paise y. Obviously the genuine mistake of the clerk at the bank counter implies both these numbers x and y are of two digits. From the conditions mentioned in the problem we can write

$$100y + x - 50 = 3(100x + y) \text{ or, } 97y - 299x = 50 \tag{i}$$

where x and y are two-digit natural numbers, i.e. $1 \leq x, y \leq 99$.

Equations like (i) with integer variables are called *Linear Diophantine Equations* in two variables. The general method of solving such equations is explained in the note at the end of this solution. At the moment, we try a simple trial and error method, which is easy to follow. In the first trial we may simply write, $y = 3x$ and $x - 50 = 3y$, assuming no carryover. Solving we get $x = -(25/4)$, which is not feasible. In the next trial we write, $y = 3x + 1$ and $x - 50 = 3y - 100$, assuming a carryover of 1 Rupee from $3y$ Paise. Now solving we get $x = (47/8)$, which is again not feasible. Next we try, $y = 3x + 2$ and $x - 50 = 3y - 200$, assuming a carryover of 2 Rupees from $3y$ Paise, when we get $x = 18$ and $y = 56$, both these numbers satisfy all the conditions mentioned earlier. So the original cheque was written for ₹ 18 and P 56.

[Note on the solution of general Diophantine equations of the form

$$ax + by = c \qquad \text{(ii)}$$

with a, b, c, x, y all integers.

First we have to check that c is divisible by the gcd (a, b). Then dividing the equation by the gcd (a, b), we rewrite the equation in the form $a_1 x + b_1 y = c_1$, where the gcd $(a_1, b_1) = 1$. Now we will solve the reduced equation of the form $a_1 x + b_1 y = 1$. With the righthand side as 1, one set of values of the unknowns (x, y) can be obtained using the Euclid's algorithm to obtain the gcd (a_1, b_1) as 1. This will be explained in detail while solving Problem No. 108 using this general method. At this stage it must be pointed out that we have to multiply the values of (x, y), so obtained by c to get to get the principal set of solution, say (x^*, y^*), for the original equation. It is imperative to mention that infinite number of solutions exist for such equations. It is easy to understand that with (x^*, y^*) as a solution, we can always write the following infinite solutions

$$x = x^* + r\frac{b}{\gcd(a, b)} \text{ and } y = y^* - r\frac{a}{\gcd(a, b)} \qquad \text{(iii)}$$

where r is an integer.]

Now we apply this general method to solve eq. (i) in Solution 108. Comparing this with eq. (ii) in the above note, and observing that the gcd (97, 299) is 1, we first solve

$$97y - 299x = 1 \qquad \text{(iv)}$$

Towards this end, we obtain the gcd (97, 299) using Euclid's algorithm as shown below:

```
97 ) 2 9 9 ( 3
     2 9 1
     ──────
      8 ) 9 7 ( 1 2
          9 6
          ───
           1 ) 8 ( 8
               8
               ─
```

The above procedure is now put in the form

$$1 = 97 - 8 \times 12$$
$$= 97 - (299 - 3 \times 97) \times 12$$
$$= 97 \times 37 - 299 \times 12$$

Comparing the above expression with (iv), it is easy to conclude $y = 37$ and $x = 12$ So for eq. (i) in Solution 108, we get $x^* = 12 \times 50 = 600$, $y^* = 37 \times 50 = 1850$.

Now using eq. (iii) we can write, infinite number of solutions as $x = x^* + 97r$, $y = y^* + 299r$, with r as an integer. In the problem both x and y are positive two digit integers, which can be obtained only with $r = -6$. Thus, we get $x = 18$, $y = 56$ as obtained earlier.

109. Let the initial number of both apples and oranges be x, which is an even number. When the cost price was recovered let the numbers of remaining oranges and apples be y and $(7 - y)$, respectively. So according to the problem, we can write the following Diophantine equation:

$$[2(x - y) + (x - 7 - y)] \times 1.1 = 3x \text{ or, } 3x - 11y = 77 \qquad \text{(i)}$$

(i) We can solve eq. (i) by trial and error remembering that $y < 8$ and x must be an even number. Trying $y = 7$ and 6, x is not an integer, $y = 5$ gives $x = 44$, which is an even number as desired. You can try with other lower values of y to convince yourself that the only feasible answer is $y = 5$, when the the net profit is ₹ $(5 \times 2.20 + 2 \times 1.10) = ₹ 13.20$.

Following the general method of solving such equations, elaborated in the note after the solution of Problem number 108, we carry out the Euclid's algorithm of determining the gcd (3, 11) as shown below:

$$3 \,) \, 1 \, 1 \, (\, 3$$
$$\underline{9}$$
$$2 \,) \, 3 \, (\, 1$$
$$\underline{2}$$
$$1 \,) \, 2 \, (\, 2$$
$$\underline{2}$$

Then we write

$$1 = 3 - 2 = 3 - (11 - 3 \times 3) = 3 \times 4 - 11 \times 1$$

Hence, the solution to the reduced equation of the form $3x - 11y = 1$, comes out as $x = 4$, $y = 1$. Thus, the principal solution of eq. (i) is $x^* = 4 \times 77 = 308$ and $y^* = 1 \times 77 = 77$. When the infinite number of solutions can be written in the general form $x = x^* - 11r$, $y = y^* - 3r$.

Now the restrictions on y is that it is less than 8, which tells that r is either 24 or 25. Now to make x an even number r must be 24 when $y = 5$. So the net profit is ₹ $(5 \times 2.20 + 2 \times 1.10) = ₹ 13.20$.

110. Since 1 cow = 1 goat + 1 rabbit, we can say 2 goats and 1 rabbit complete the job in 45 days. Furthermore, it is given that 1 goat and 1 rabbit finish the job in 90 days, so we can say that 1 goat finishes the initial amount of grass present in the field (say this amount is G) in 90 days and 1 rabbit finishes

the daily growth. So on each day, the cow and the goat eat, respectively (1/60) G and (1/90) G amount of grass.

Hence, they together consume $[(1/60) + (1/90)]$ G $= (1/36)$ G amount of grass and 1 rabbit eats up the daily growth.

Hence, the three together will make the field devoid of grass in 36 days.

111. Let us imagine 13 places in a line, in three of which we can keep three sticks, numbered say 1, 2 and 3 and in the remaining 10 places we can keep the balls. On the left of the first stick we keep only red balls, between the first and the second stick we keep only blue balls, between the second and the third stick we keep only yellow balls and on the right of the third stick we keep only green balls. In this way, we have created four compartments for the balls of different colours. Thus, the positioning of the sticks defines an arrangement. Each separate arrangement defines the content of a different pack. So the number of different arrangements, i.e. the number of possible different packs is equal to the number of different ways three sticks can be kept in 13 places. Thus, the answer to the problem is

$$\binom{13}{3} = {}^{13}C_3 = \frac{13.12.11}{3.2} = 286$$

112. Using the conditions given in the problem we make the following list:

Position in the line k	The fraction of cake received C_k	The fraction of cake remaining
1	(1/100)	(99/100)
2	(2/100)(99/100)	(99/100)(98/100)
3	(3/100)(99/100)(98/100)	(99/100)(98/100)(97/100)
⋮	⋮	⋮
k	C_k	
$k+1$	C_{k+1}	

Observing the entry for $k = 2$ and 3, it is easy to write

$$\frac{C_k}{C_{k+1}} = \frac{k}{k+1} \frac{100}{100-k}$$

and also note that $C_2 > C_1$.

So initially the amount of cake received is increasing, hence the maximum amount is received by the kth person if $C_k > C_{k+1}$.

Thus, for

$$\frac{C_k}{C_{k+1}} > 1, \text{ or, } \frac{k}{k+1} \frac{100}{100-k} > 1$$

we get $100k > (k + 1)(100 - k)$ or, $k^2 + k > 100$.

This condition is first satisfied when $k = 10$.

Hence, to receive the largest piece, one should stand in the 10th position.

113. For the nth person in the queue to get a free ticket, $(n - 1)$ persons in front of him must have different birthdays and his birthday must be the same as that of one of these $(n - 1)$ persons.

The probability of this event can easily be written as

$$P(n) = \left[\frac{364}{365} \times \frac{363}{365} \times \frac{362}{365} \times \cdots \times \frac{365 - n + 2}{365} \right] \times \frac{n - 1}{365} \quad \text{(i)}$$

For this probability to be maximum

$$P(n) > P(n + 1) \text{ or, } \frac{P(n)}{P(n + 1)} > 1$$

Using eq. (i), we get

$$\frac{365}{365 - n} \times \frac{n - 1}{n} > 1 \text{ or, } n^2 - n - 365 > 0$$

For a natural number n, the above condition is first satisfied when $n = 20$. Hence, to get the free ticket, the person should stand at the 20th position in the queue.

114. Let the volume of the container be V litres. After removal of milk for the first time, the amount of milk in the container is $V_1 = (V - 3)$ litres. After replenishing with water the amount of milk per litre of the mixture is $S_1 = (V - 3)/V$ litre. After removal of milk for the second time, the amount of milk left in the container is $V_2 = (V - 3 - 3S_1)$. After replenishing with water, the amount of milk per litre of the mixture is

$$S_2 = \frac{V_2}{V} = \frac{\left(V - 3 - 3\frac{V-3}{V} \right)}{V}$$

After removal of milk for the third time the amount of milk in the container is $V_3 = V_2 - 3S_2$. Again after replenishing with water, the amount of milk per litre of the mixture is $S_3 = V_3/V$. Now after substituting for V_3 and S_2, we get

$$S_3 = \frac{(V - 3)^3}{V^3} = \left(1 - \frac{3}{V} \right)^3$$

Equating this value of S_3 to $(1/2)$ as given in the problem, we obtain

$$1 - \frac{3}{V} = \left(\frac{1}{2} \right)^{1/3} \approx 0.7937$$

$$\text{or, } V \approx \frac{3}{(1 - 0.7937)} = \frac{3}{0.2063} \approx 14.542 \text{ litres}$$

115. The first convict is given a mixture of a drink by taking a drop from each of the first 500 bottles. If he dies, then the second convict is given a similar mixture from the first 250 bottles. By continuing this process, in 10 steps we can complete inspection of 2^{10} = 1024 bottles and detect the spiked bottle. Needless to say, at any stage if the convict survives then the other half of the bottles should be taken and tested similarly. At any stage when an odd number, say $2m + 1$, bottles remain then in the next stage $(m + 1)$ bottles can be considered for making the mixture.

116. This time the minister marks the 1000 bottles with labels 000 to 999. These numbers are then expressed in binary notation (base 2) as shown in the Table below:

Table 5.1

Bottle number expressed in decimal notation	Bottle number expressed in binary notation
000	0000000000
001	0000000001
002	0000000010
003	0000000011
$\vdots$	$\vdots$
999	1111100111

The minister now onward considers the bottle numbers expressed in the binary notation. First he makes a mixture by taking a drop from each of the bottle with a number having 1 at the right most position. Then the second mixture is made by taking a drop from each of the bottle with a number having 1 at the second position (from the right end). This way he makes 10 mixtures as there are 10 positions in the binary expressions. Then the kth convict is given the kth mixture with $k = 1, 2, 3, \ldots, 10$. On the 30th day if the convicts who had mixture numbers say $k = p, q, r$, etc. then the spiked bottle number, expressed in decimal notation will be $2^{p-1} + 2^{q-1} + 2^{r-1}$. And if all the 10 convicts survive then the bottle number 000 is the spiked one, which was not used in any mixture.

117. We approach the problem in an indirect manner. We say that with k drop-pings, the highest floor of the building that can be determined is given by $M(k)$. Towards this end, the first egg is dropped from the kth floor. If it breaks then starting from the first floor, the second egg is dropped from each successive floor (up to the $(k - 1)$th) is tested until it breaks. From which the value of $M(k)$ can be determined. If the first egg does not break in the first dropping from the kth floor, then this first egg is taken to $k + (k - 1) = (2k - 1)$th floor and dropped. This is the second dropping.

If it breaks, then starting from $(k + 1)$th floor each successive higher floor (up to the $(2k - 2)$th floor) is tested. So with k droppings we can cover $k + (k - 1) = (2k - 1)$th floor. If the egg does not break when dropped from the $(2k - 1)$th floor, then we go to the $k + (k - 1) + (k - 2)$th $(= (3k - 3)$th) floor and make the third dropping. Continuing this way with total k droppings we can cover up to the maximum floor level given by

$$M(k) = k + (k - 1) + (k - 2) + \cdots + 3 + 2 + 1 = \frac{k(k + 1)}{2}$$

For testing a 100-storied building, the minimum value of k will be given by $M(k) \geq 100$, or the lowest value of k is 14.

For example, by dropping 14 times successively from 14, 27, 39, 50, 60, 69, 77, 84, 90, 95, 99 and 100th floor, we can determine the highest floor level (whatever may be that value), of a 100-storied building, dropped from which the egg breaks. Note that this not a unique solution of dropping sequence, In this sequence, the best case scenario is when the egg breaks from the 100th floor in the 12th dropping and in the worst case scenario, it breaks from the 13th floor in the 14th dropping. If the egg breaks from 26th floor, then also we need 14 droppings and so on.

118. Let the minimum number of coconuts collected initially be N and the first to the fifth person had hid, respectively a, b, c, d and e number of coconuts with them in their night raids to the heap. Then we can write

$$N = 5a + 1 \text{ or, } N + 4 = 5(a + 1)$$
$$\text{and } 4a = 5b + 1 \text{ or, } 4(a + 1) = 5(b + 1)$$

Proceeding in a similar manner, it is easy to write considering all the way up to the fifth person the following equations:

$$4(b + 1) = 5(c + 1)$$
$$4(c + 1) = 5(d + 1)$$
$$4(d + 1) = 5(e + 1)$$

Using all the above equations (substituting backwards), we finally get

$$N + 4 = 5 \times \left(\frac{5}{4}\right)^4 (e + 1)$$

For minimum integer value of N, $(e + 1) = 4^4 = 256$, when $N = 3525 - 4 = 3121$ and $e = 256 - 1 = 255$. It must be noted that the number of coconuts left in the next morning is $4 \times 255 = 1020$ which is divisible by 5 as desired.

So the answer is that the minimum value of the initial number of coconuts is 3121.

Alternative way: It may be noted that dividing −4 (negative 4) by 5 we may consider the quotient is −1 when the remainder is (−4 − (−5)) = 1. Taking away this quotient and remainder from −4 we are not taking away anything (as −1 + 1 = 0), so again we get back −4.

Hence, if we start, in an abstract sense, with an initial −4 number of co-conuts, then we can continue to give −1 to the man making the night raid to the heap and throw away +1 remainder to the monkey forever. So to divide by 5 consecutively 5 times, we can start with 5^5 numbers of coconuts for the persons to get their share along with −1 they get from −4, we can satisfy all the conditions. Thus, to start with the minimum number of co-conuts required is $5^5 - 4 = 3121$, as obtained before. One needs to check that with this starting value, finally 1020 coconuts are left for the next morning and this number is divisible by 5 as desired.

119. Following the first method of solution for Problem number 118, we can easily write, the minimum value of the initial number of coconuts is

$$N + 4 = \frac{5^6}{4^5}(f + 1)$$

where f is the number of coconuts everybody gets on the next day morning.

For minimum integer value of N, $(f + 1) = 4^5$, when $N = 5^6 - 4 = 15{,}621$.

120. After opening one door showing a goat, it is clear that the car is behind one of the two closed doors. So each door has probability (1/2) of hiding the car, so no need to change the decision. But that conclusion is wrong. To decide correctly let us consider all the possibilities.

Without any loss of generality we number the initially chosen door as 1 and the others as 2 and 3. The following Table shows all the three possibilities with this numbering system.

Door 1	Door 2	Door 3
Car	Goat	Goat
Goat	Car	Goat
Goat	Goat	Car

So out of three possible situations changing the choice will be wrong only in the first case, when either door 2 or 3 can be opened to show a goat. But in the second case, door 3 will be opened to show a goat changing the choice from door 1 to 2 is desired. In case 3, door 2 will be opened to show a goat and again changing the choice from door 1 to 3 is desired. Thus, in two out of possible 3 situations a change desirable.

Hence, the probability of winning the car will be 2/3 (> 1/2) by changing the initial choice after one door is opened to show a goat.

121. We must note the following points:

(i) So long there are more than 2000 bananas at the starting point, the person has to return to the starting point twice. But if the number of bananas at the starting point is more than 1000, but less than or equal to 2000, then one has to return to the starting point once.

(ii) For most efficient transport arrangement we should load the camel as much as possible.

(iii) From the above two points we conclude that at the last step one should start travelling with 1000 bananas towards the destination, so that there is no need to return any more.

It may be noted that so long one has to return twice to the starting point, the camel consumes 5 bananas to cover a distance of 1 km; three bananas for three forward trips and two bananas for two return trips. So if we travel 200 km this way 1000 bananas are consumed by the camel and 2000 bananas remain at this intermediate point (say A), at a distance of 200 km from the original starting end and 800 km from the destination. For example, in the first trip starting with 1000 bananas, travel 200 km and dump 600 bananas and start returning with 200 bananas and reach the original starting point with no banana. Then a similar second trip is made with 1000 bananas and after dumping 600 bananas at A return to the starting point, load 1000 bananas and reach the point A with 800 bananas. Thus, 2000 bananas reach the point A. In the second phase starting from the point A make two trips, each starting with 1000 bananas and travelling 333 km to reach another location say B at a distance $1000 - (200 + 333) = 467$ km from the destination. In these trips 3 bananas per km of travel are consumed by the camel, so in total 999 bananas are consumed and 1001 bananas and the reach the point B. Leaving one banana there (or eating it if the person likes) the camel is loaded with 1000 bananas and reach the destination after travelling the remaining 467 km; the camel consumes 467 bananas during this travel.

Hence, $1000 - 467 = 533$ bananas reach the destination. This is the maximum, you may try any other way to convince yourself that there is no way you can improve upon this figure.

122. Observing Fig. 4.1, we can say that the sum of the diameters of all the (infinite number of) circles is equal to the height of the height (h) of the

isosceles triangle. It is easy to find that $h = \sqrt{13^2 - 5^2} = 12$. So the sum of the circumferences of all the circles $= 12\pi$.

123. Let the lengths of the three sides of the triangle be x, x and y. We know from triangle inequality $2x > y$. Since the length of all sides must be less than or equal to 1994, we consider two different cases:

(i) $1 \le x \le \frac{1994}{2}$, i.e. $1 \le x \le 997$

(ii) $998 \le x \le 1994$.

For case (i), for any value of x, the value of y can be any integer from 1 to $(2x - 1)$. Thus, the possible number of triangles in this category is

$$N_1 = \sum_{r=1}^{997}(2x - 1) = (997)^2$$

(Recall that the sum of consecutive odd numbers is a perfect square.)

For case (ii), for each of the 997 values of x, the value of y can be any integer from 1 to 1994.

Hence, the total number of triangles in this category is $N_2 = 997 \times 1994 = 2(997)^2$.

Hence, the total number of triangles is

$$N_1 + N_2 = 3 \times (997)^2 = 3 \times (10^3 - 3)^2 = 3 \times (10^6 - 6 \times 10^3 + 27) = 2982027$$

124. Let the three angles of the triangle be given by integers x, y and z, all expressed in degrees. It is known that $x + y + z = 180$.

For determining all possible values of x, y and z, we consider all integers from 1 to 180 are marked on the (real) number line. There are 179 gaps between these 180 numbers. Placing 2 sticks in these 179 gaps we can divide the number 180 in three integer parts. Let us call the number immediately to the left of the first stick x. The number of integers falling between the two sticks is called y and the number of integers lying to the right of the second stick is called z. It is easy to see that two sticks can be placed in 179 gaps in

$$\binom{179}{2} = \frac{179 \times 178}{2} = 179 \times 89$$

different ways. At this stage it may be pointed out that triangles are three types, namely,

(i) scalene with all three angles different

(ii) isosceles with two angles of same measure (with three angles as x, x and $180 - x$)

(iii) equilateral with all three angles equal to $60°$.

Now we discuss the issue of similar triangles. The naming x, y and z was arbitrary we could have called the *three sequential parts* as y, x and z or as z, x and y and so on. Thus, six similar scalene triangles are with angles (x, y, z), (x, z, y), (y, x, z), (y, z, x), (z, x, y) and (z, y, x), where the three numbers are representing, respectively, the angles A, B and C of the triangle ABC.

Similarly, three similar isosceles triangles will have angles $(x, x, 180 - x)$, $(x, 180 - x, x)$, $(180 - x, x, x)$, where x can take 88 different values, viz. 1, 2, 3, ..., 59, 61, 62, ..., 89. Note that the value 60 is missing in this list as that will give an equilateral triangle.

Hence, finally we get the number of dissimilar scalene triangles

$$N_1 = \tfrac{1}{6}(179 \times 89 - 3 \times 88 - 1) = 2{,}611$$

Similarly, the number of dissimilar isosceles triangles is 88 and that of equilateral triangle is 1. So the total number of dissimilar triangles with integer angle measures is $2{,}611 + 88 + 1 = 2{,}700$.

125. It may be noted that in each step as the value of k increases by 1, the perimeter of the figure increases by a factor $(4/3)$. Thus, after n steps the perimeter of the bounding curve is $3 \times (4/3)^n$. so as $n \to \infty$, the perimeter also tends to infinity. But the area enclosed, A, by this bounding curve of infinite length remains finite as calculated below.

$$
\begin{aligned}
A &= \frac{\sqrt{3}}{4} + 3 \times \left(\frac{\sqrt{3}}{4} \times \frac{1}{9} \right) + 12 \times \left(\frac{\sqrt{3}}{4} \times \frac{1}{81} \right) + \ldots \\
&= \frac{3\sqrt{3}}{4 \times 4} \times \left(\frac{4}{3} + \frac{4}{9} + \frac{4^2}{9^2} + \ldots \right) \\
&= \frac{3\sqrt{3}}{16} \times \left(\frac{1}{3} + 1 + \frac{4}{9} + \frac{4^2}{9^2} + \ldots \right) \\
&= \frac{3\sqrt{3}}{16} \left(\frac{1}{3} + \frac{1}{1 - \frac{4}{9}} \right) = \frac{3\sqrt{3}}{16} \left(\frac{1}{3} + \frac{9}{5} \right) = \frac{2}{5}\sqrt{3}
\end{aligned}
$$

126. $N = (1!)(2!)(3!) \ldots (100!)$

   $= 2(1!)^2 . 4(3!)^2 6(5!)^2 \ldots 100(99!)^2$

   $= 2^{50}(1.2.3 \ldots 50)(1!)^2 (3!)^2 (5!)^2 \ldots (99!)^2$

   $= (2^{25})^2 (50!)(1!)^2 (3!)^2 (5!)^2 \ldots (99!)^2$

   Hence, $N/50!$ is a perfect square.

   You can easily generalize this to $N = (1!)(2!)(3!) \ldots (2n!)(2n+1)! \ldots (4n)!$ to be divided by $(2n!)$ to render a perfect square.

127. The positive integers 1 to 2^4 are to be split maintaining the sum, the sum of squares and the sum of cubes the same. So let us start with trying to split

integers 1 to 2^2 into two groups maintain only the sum same. This is easily seen as (1, 4) and (2, 3). Next we try to split positive integers 1 to 2^3, into two groups so that the sum and the sum of the squares remain the same for the two groups. In the new group (5, 6, 7 and 8) we can easily see (5, 8) and (6, 7) have the same sum. If we crosslink with the previous grouping as (1, 4, 6 and 7) and (2, 3, 5 and 8), then these two groups are seen to have both the sum and the sum of the squares identical. Also note that we add $2^2 = 4$, to the original elements and then crosslink the two groups. This means the second grouping is done as $(1, 4, 2 + 2^2, 3 + 2^2)$ and $(2, 3, 1 + 2^2, 4 + 2^2)$. Now by induction we can prove that if integers from 1 to 2^m can be partitioned into two groups with the sum of powers $(m - 1)$, then integers 1 to 2^{m+1} can be portioned into two groups with all the sums up to the powers m of their elements, by cross-linking terms with additions of 2^m. So we add $2^3 = 8$ with previous groups of 4 in the crossed manner to obtain (1, 4, 6, 7, 10, 11, 13 and 16) and (2, 3, 5, 8, 9, 12, 14 and 15). You can verify that these two groups have the same sum, the same sum of squares and the same sum of cubes.

Proof by induction: we have already seen that integers 1 to 2^m can be portioned into two equal sized disjointed sets X and Y of $\Sigma\{P(x) : x \in X\}$ is equal to $P(Y)$ having the same sum of up to the jth powers with j running from 0 to $(m - 1)$. For moving up to 2^{m+1} level we take a new set defined as $X' = X \cup (Y + 2^m)$ and $Y' = Y \cup (X + 2^m)$. We need to prove that that for any polynomial P of degree at most m

$$P(X) + P(Y + 2^m) = P(Y) + P(X + 2^m) \tag{a}$$

If P has degree less than m, then by induction $P(X) = P(Y)$ and we also get $P(X + 2^m) = P(Y + 2^m)$. Thus, X' and Y' certainly agree for polynomials of order less than m. But even for the order m, it is alright as the both sides of eq. (a) are the same as the mth power terms of $P(x) + P(Y)$. It is obvious that this kind of partition can be done for 1 to 2^n, for all values of $n > 1$ for sums up to the powers $(n - 1)$.

128. The first 11 digit binary number is $2^{10} = 1024$. So up to 1000 can be covered by maximum ten-digit binary numbers. So out of these 10 at most six can be 1 and the rest are zeros. We make the following list of total numbers that can be written using different numbers of 1's.

Using one 1 = $\binom{10}{1} = {}^{10}C_1 = 10$ Using two 1's = $\binom{10}{2} = {}^{10}C_2 = 45$

Using three 1's = $\binom{10}{3} = {}^{10}C_3 = 120$ Using four 1's = $\binom{10}{4} = {}^{10}C_4 = 210$

Using five 1's = $\binom{10}{5} = {}^{10}C_5 = 252$ Using six 1's = $\binom{10}{6} = {}^{10}C_6 = 210.$

Adding all these numbers, we get 847. But two of these numbers like 1111110000 and 111110100 are 1008 and 1000, respectively. So $847 - 2 = 845$ numbers all less than 1000 can be written using six or fewer 1's.

129. With n = even the given number is even and greater than 2 hence cannot be a prime. With n = odd (and > 1), let $n = 2k + 1$, with $k \geq 1$.

Now we can write the given number

$$N = (2k + 1)^4 + 4^{2k+1}$$
$$= (2k + 1)^4 + 4 \cdot 4^{2k}$$
$$= (2k + 1)^4 + 4 \cdot (2^k)^4$$

Now we substitute $a = 2k + 1$ and $b = 2^k$, when $N = a^4 + 4b^4$, which can always be factorized as shown below:

$$a^4 + 4b^4 = (a^2)^2 + 4a^2b^2 + 4b^4 - 4a^2b^2$$
$$= (a^2 + 2b^2)^2 - (2ab)^2$$
$$= (a^2 + 2b^2 + 2ab)(a^2 + 2b^2 - 2ab)$$

This identity is known as *Sophie Germain's identity.*

Hence the given number N cannot be a prime.

130. (a) We note that $(\sqrt{2} + \sqrt{3})^2 = 5 + 2\sqrt{6}$ and $(\sqrt{3} - \sqrt{2})^2 = 5 - 2\sqrt{6}$. Similarly, $(\sqrt{2} + \sqrt{3})^4 = 49 + 20\sqrt{6}$ and $(\sqrt{3} - \sqrt{2})^4 = 49 - 20\sqrt{6}$.

Adding we notice that $(\sqrt{2} + \sqrt{3})^{2n} + (\sqrt{3} - \sqrt{2})^{2n}$ is always an integer. Since for large values of n, such as 50, the value of $(\sqrt{3} - \sqrt{2})^{100}$ will be a very small number as $\sqrt{3} - \sqrt{2} \approx 0.318$.

Hence, we conclude $(\sqrt{2} + \sqrt{3})^{100}$ is almost an integer so the digit after the decimal point must be 9.

(b) Let $\sqrt{181 + 12\sqrt{5}} = \sqrt{a} + \sqrt{b}$. Squaring both sides and equating the rational and irrational parts we obtain $181 = a + b$ and $720 = 4ab$ or, $a + b = 181$ and $ab = 180$, when it is readily seen that a and b are 1 and 180 or vice versa. We again write this number as $\sqrt{m} + \sqrt{n}$.

Thus, the given number was $\sqrt{13 + 1 + \sqrt{180}}$. We again write this number as $\sqrt{m} + \sqrt{n}$. Squaring we get $m + n = 14$ and $mn = 45$.

Hence, m and n are obtained as 9 and 5 or vice versa.

Thus, the given number is $\sqrt{9} + \sqrt{5} = 3 + \sqrt{5}$.

131. Let the two digits in base 11 be x and y. So it is given $11x + y = 16y + x$. So $10x = 15y$ or, $2x = 3y$. To get a three digit number in base 10, y must be at least 6 when $x = 9$. Thus, the smallest two-digit number is $16.6 + 9 = 105$.

132. Let the prime numbers be p (for A) and q (for B). It is given, $p^2 - q^2 - 8p - 12q = 20$. Or, $(p-4)^2 - (q+6)^2 = 0$ or, $(p-4) = (q+6)$, as $p > q$, or $p-q = 10$. Two largest two-digit primes differing by 10 are 79 and 89. So the largest possible sum is $79 + 89 = 168$.

133. Every sum can be written as $1a + 3b + 9c + 27d$, where a, b, c, d can be either $-1, 0$ or 1.

Adding $1+3+9+27 = 40$ to these sums, we get a, b, c and d as $0, 1$ or 2. Thus, adding 40 to all possible sums, we get all the four-digit numbers with base 3. There are $3^4 = 81$ such numbers. Now subtracting 40 and discarding 0, the number of different positive sums is 40.

134. Let us define the operation of erasing and substituting two positive integers m and n as $m \otimes n = m + n + mn$.

First we observe that this operation is associative, i.e. $m \otimes (n \otimes p) = (m \otimes n) \otimes p$ as shown below:

The lefthand side is $m \otimes (n + p + np) = m + n + p + np + mn + mp + mnp$ and the righthand side is $(m + n + mn) \otimes p = m + n + mn + p + mp + np + mnp$ and both are same.

Thus, the process is independent of order and we start from the smallest number and proceed to get 1 and 2 is replaced by $1 + 2 + 1 \times 2 = 5 = 3! - 1$. Next erasing this number and 3, we write $5 + 3 + 5 \times 3 = 23 = 4! - 1$. Continuing the same way erasing this number and 4, we write $23 + 4 + 23 \times 4 = 119 = 5! - 1$. So the pattern is clear the last remaining number will be $101! - 1$. You may convince by noting that in general $m + n + mn = m + n + mn + 1 - 1 = (m + 1)(n + 1) - 1$.

135. The given number is written as

$$\sqrt{(2021^2 - 2 \times 2021 + 2)(2021^2 + 2 \times 2021 + 2) - 4}$$
$$= \sqrt{(2021^2 + 2)^2 - 4 \times 2021^2 - 4}$$
$$= \sqrt{(2021^2)^2} = 2021^2 = 4084441$$

So the sum of the digits is 25.

136. It is easy to see by Pythagoras's Theorem, that the final number will be

$$\sum_{n=1}^{100} \sqrt{n^2} = \sqrt{\frac{100 \times 101 \times 201}{6}} = \sqrt{338,350}$$

(See Appendix B)

137. Let us start from the last step when one pebble must have been taken from a pile of two pebbles to create the 100th pile. So at this stage you note the product as $1 \times 1 = 1$. Now we see that with 2 pebbles to start with the

sum of the product to get two piles of one pebble each is 1. Now consider 3 pebbles to start with. In the first step, we get the product of the number of pebbles in the two freshly created piles is $1 \times 2 = 2$.

Thus, the sum of the products $1 + 2 = 3$. Proceeding this way starting with 4 pebbles one gets $1 + 2 + 3 = 6$. So starting with 100 pebbles the sum of the products will be

$$\sum_{1}^{99} n = \left(\frac{99 \times 100}{2} \right) = 4{,}950$$

Chapter 6

Computer and programming fundamentals

We are now in a position to appreciate the potential of different types of numbers and their applications in describing real-life problems and their solutions. It is also expected that you might be eager to use a computer for solving some problems. Computers would be handy not only to save time, but to relieve you from the drudgery of manual computation as well. Last but not the least, computers may be a big help if you could not obtain an analytical solution.

Here we will learn about the following, with the final goal of solving some of the problems presented in this book without the rigorous analytical acumen required otherwise.

- How the numbers and codes are represented in computers.
- Logic operations and logic gates.
- How arithmetic and logic operations are done in a computer.
- A Programming language with a tutorial introduction to programming in a concise form.

Well, it seems to be a tall-order; believe us it would be very short and fun to learn.

6.1 Advantages of binary representation of numbers

In a computer the numbers are represented in binary. In fact, not only the numbers, the instruction or any other information is nothing other than some 0's and 1's stored in the memory, the main resource of the computer. A switch with its two states, namely, OFF and ON is the simplest possible 2-state device that be used to represent 0 and 1. So, a few switches together, say 8, may be used to represent a 8-bit binary number $(b_7 b_6, \ldots, b_0)$ 00000000_2 to 11111111_2; i.e. 0 to 255 in decimal. Additionally, the ON and OFF states of the switches can be mapped to

TRUE and FALSE in the logical domain. This is very important as the computer relies on logic operations as we shall see next.

In short, if we choose to represent a number in decimal form, the computing system should be able to represent and identify 10 different states; 0 to 9 and carry out all arithmetic operations using 10-state devices. This leads to a complicated, costly and bulky solution. World's first electronic computer ENIAC (1946), designed and installed in the University of Pennsylvania, USA used decimal representation of numbers. Moreover, transistors and ICs were not invented then and it was built using electronic valves. ENIAC weighed 27 tons, occupied 18,000 sq. ft. and consumed ONLY 150 KW of electricity.

Note that in electric/electronic circuits the logic states are actually some voltage (or current or direction of magnetism for magnetic devices like a disk). We may assume that the 10 decimal digits would be represented by 10 different voltages, say $0, 1, \ldots, 9$ and in case of binary digits they are only 0 and 9 volt representing binary digits 0 (FALSE) and 1 (TRUE), respectively. This arrangement is shown in Fig. 6.1(a) that clearly exhibits the advantage of working with fewer number of digits (or states).

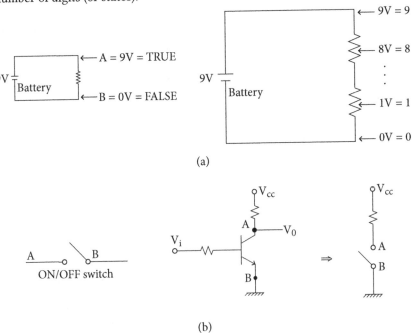

Fig. 6.1: (a) Simulating two states and ten states, (b) Transistor as a switch

The circuitry becomes simple and less costly if we have appropriate devices to identify and produce these 2-states. In reality, we use transistors that can be turned on and off very easily. So, a transistor can act as a switch in logic circuitry representing perfectly the *binary* digi*t* or the *bit* as shown in Fig. 6.1(b). With a low in V_i the transistor does not conduct and V_o is high. Conversely, with a high V_i the transistor conducts making V_o low.

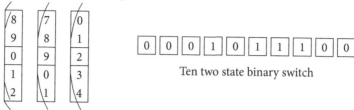

Ten two state binary switch

Three ten position rotary switch

Fig. 6.2: A 10-state rotary switch versus 2-state on-off switch based odometer

The reduction from 10-state to a 2-state device for simplicity and cost reduction in implementing logic circuits is shown in Fig. 6.2 in the context of an odometer in a car with 3 rotary wheels.

The 3-rotary wheels of the odometer are marked from 0 to 9. Together they represent, say 092 indicating that the car has travelled 92 KM. A simpler arrangement might be ten ON-OFF switches marked 0 (OFF) and 1 (ON) and placed side by side. These switches form a 10-bit binary number; and the reading 00 0101 1100 would mean the car has travelled a total of 92 KM so far. So, three 10-state rotary switch can be replaced by ten, much simpler 2-state ON-OFF switch to represent numbers. The only disadvantage is, perhaps, we need switches in greater numbers.

However, the modern Integrated Circuit (IC) technology allows us to integrate Billions (or even more) transistors in the area of a finger nail ($\approx$ 10 mm $\times$ 10 mm) along with their interconnections practically at no cost. This is a convincing reality and for the same reason numbers or other piece of information (like a text) are represented in *binary* for all sorts of electronic devices including computers. So, simplicity, cost and space reduction can easily be achieved.

The reader now, we hope, gets the idea why the binary number system is a universal choice for computers. This transition from 10 different digits (or symbols) to 2 digits reduces the circuit complexity and a perfect natural mapping with the two logic states (FALSE and TRUE). Thus, the use of the binary number system becomes universal in any computer; supercomputers, to hand held devices like your mobile phone.

Octal and hexadecimal representations

Binary bit strings are often 16 bit or 32 bit (or even 64 bit) long. It is error prone and time consuming to write them in the binary form by humans. Octal (base = 8; and the digits 0, 1,..., 7) or hexadecimal (base=16 and the digits 0, 1, ..., 9, A, B, C, ..., F) representations help reduce these long bit strings and make them manageable. For example, a 16-bit binary number 10|110|011|010|010|101|110 (vertical lines are placed after every 3rd bit starting from the rightmost position for a better readability and easier conversion) can be written in octal as 2632256_8. The same binary number 1011|0011|0100|1010|1110 in hexadecimal is $B34E_{16}$. See Appendix F for a table showing decimal number 0 to 15 and their equivalent binary, octal and hexadecimal (Hex) representations.

We may discuss next how a computer using numbers in binary form can carry out the basic arithmetic operations; at least addition. Dropping all other issues we focus on the core part of the computer that carries out addition.

6.2 A computer as a black box

Confronted with the challenge of building at least the core arithmetic part of the computer let us make a modest beginning. Consider a computer as a black box (see Fig. 6.3). It receives binary input and produces binary outputs and could carry out addition only. Now we may focus on how addition is carried out, to figure out the logic circuits that will be present inside the black box.

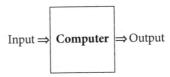

Input $\Rightarrow$ **Computer** $\Rightarrow$ Output

Fig. 6.3: Computer as a black box

How do we add?

The numbers will be represented in binary; i.e. finite number of bits together in the form of a bit-string. Thus, an n-bit number, say A, in binary would be represented as $A_{n-1}A_{n-2}\ldots A_1A_0$. To make the matter simple consider the problem of adding two bits at a time. Here we have only four input combinations in binary instead of 100 different combinations needed for the decimal.

The addition logic in a tabular form, the Half-adder (HA) truth-table to be precise, is presented in Fig. 6.4.

Input		Output	
A	B	C	S
0	0	0	0
0	1	0	1
1	0	0	1
1	1	1	0

Fig. 6.4: Truth-table for a half-adder.

Here, A and B represent the bits to be added (and not the n-bit string) while the carry (C) and the sum (S) together represent the outcome. Note that considering the inputs as binary digits (i.e. 0 and 1) we see that the truth-table complies with the addition of two 1-bit binary numbers.

Before we construct a half-adder or other logic circuits introduction to logic operations and logic gates is required and is discussed next.

6.3   Logic operation and logic gates

Logic circuits based on logic operations are the basic building blocks in a computer. The three primitive logic operations are (i) NOT, (ii) AND, and (iii) OR, the corresponding implementations, usually using electronic circuitry are known as *logic gates*. We are not actually bothered with the exact implementation of the logic circuitry and use only the logic gate symbols to avoid complexity.

6.3.1   NOT, AND and OR logic

Fig. 6.5 shows the truth-table, logic symbol, simulation of the circuitry, etc. for all the primitive or basic logic operations.

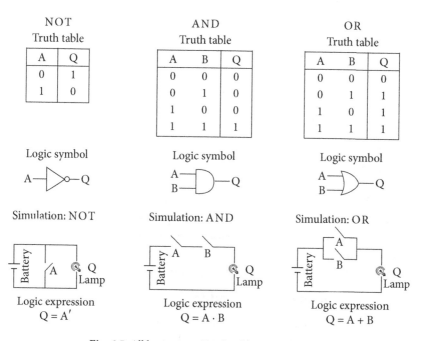

Fig. 6.5: All basic gates: Truth-table, symbol, etc.

Do keep in mind that in actual implementation transistors as switches are used to implement the logic gates. For example, in Fig. 6.1(a) consider A as the input and Q as the output then it follows NOT logic; i.e. A = 0 makes Q = 1 and A = 1 makes Q = 0.

6.3.2 The derived logic operations and logic gates

We have some derived logic operations and logic gates as well. The derived gates are mostly used to implement adders, subtracters and various other combinational circuits present in any digital system including computers. The truth-table of the derived logic operations, namely, NAND, NOR and XOR are shown in Fig. 6.6. NAND and NOR are known as *Universal gates* as using them we could make other basic gates like NOT, AND and OR. The XOR has special use in controlled inverter as well as comparator.

Name, logic symbols		Truth table		
Operation	Logic expression	Input (A and B)		Output Q
NAND $\qquad$ A⟩—▷○-Q $\qquad$ A⟩○-Q	$(A \cdot B)'$	0 0 1 1	0 1 0 1	1 1 1 0
NOR $\qquad$ A⟩—▷○-Q $\qquad$ A⟩○-Q	$(A + B)'$	0 0 1 1	0 1 0 1	1 0 0 0
XOR $\qquad$ A B A B $\qquad$ A⟩○-Q	$A \cdot B' + A' \cdot B$	0 0 1 1	0 1 0 1	0 1 1 0

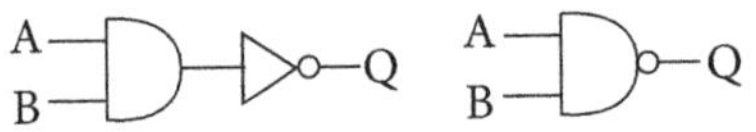

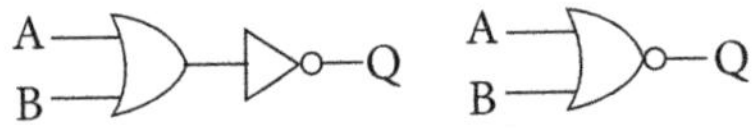

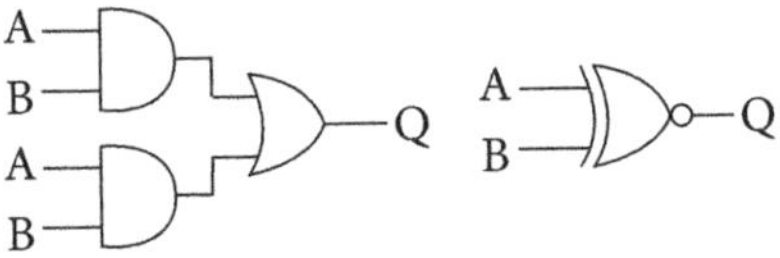

Fig. 6.6: N A N D, N O R, XOR operations, logic symbols and truth tables

This completes our discussion on the logic operations and logic gates. The logic operations follow Boolean Algebra (A tribute to the English Mathematician George Boole).

See Appendix F for the list of Boolean algebraic rules.

6.4 Implementing the half-adder

Now our patience would reap the dividend and we can implement the HALF-ADDER (HA) circuit with the help of logic gates as the building blocks.

6.4.1 The Half-adder (HA) using simple logic gates

From the HA truth-table (Fig. 6.4) we see that the CARRY output from the truth table of the HA is TRUE (1) iff (if and only if) both the input bits are 1. So, a simple two-input AND gate would do the job. The SUM is 1 whenever one input bit is

1 and the other is 0; complying the two-input XOR logic. Thus, HA logic circuit comprises of a two-input AND gate and a two-input XOR gate. Notationally, S = A · B′ + A′ · B = A ⊕ B and C = A.B. The HA as a building block, truth-table and the logic gate implementation are shown in Fig. 6.7.

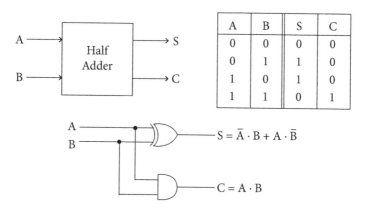

Fig. 6.7: Half-adder building block, truth-table and logic circuit implementation

6.4.2 Half-adder is not enough for addition

The deficiency of the HAs for addition can be seen easily if we try to add two n-bit ($n = 4$ in this case) numbers, say A ($a_3 a_2 a_1 a_0$) and B ($b_3 b_2 b_1 b_0$) as shown in Fig. 6.8.

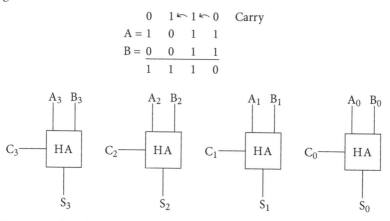

Fig. 6.8: A 4-bit addition showing the carry bits and four HA's to do the addition

Here for 4-bit operands four half-adders are employed which may take care of addend and augend bits; however, there is no provision to accommodate the third input bit which is the carry bit coming from the previous position.

6.4.3 Full-adder (FA)

Thus, in a FA there would be three inputs; the bits of addend and augend plus any carry bit from the lower to the next higher position. The truth table and the FA as a black-box are shown in Fig. 6.9.

A_n	B_n	C_{n-1}	C_n	S_n
0	0	0	0	0
0	0	1	0	1
0	1	0	0	1
0	1	1	1	0
1	0	0	0	1
1	0	1	1	0
1	1	0	1	0
1	1	1	1	1

Fig. 6.9: Full-adder truth-table and the black-box representation

Analysis of the truth table leads to the following equations for the Sum as $S_n = A_n \oplus B_n \oplus C_n$ and the Carry $C_{n+1} = (A_n \oplus B_n) \cdot C_n + A_n \cdot B_n$. Fig. 6.10 shows FA implementation using two HAs' and one OR gate and the use of four FAs' to add two 4-bit numbers.

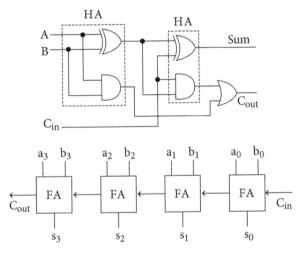

Fig. 6.10: Up: FA using HA logic; Below: 4-bit addition using four FAs'

So, the stage is set; make available the numbers in the binary form and add them using full-adders. In a modern computer the adder can add two 64-bit operands[1]. Thus, the corresponding circuitry would be an ensemble of 64 such full-adders in tandem. The number of transistors required to implement a 64-bit full adder is $64 \times 9 \times 4 = 576$. Note that 9 NAND gates are required for an FA and 4 transistors per NAND gate. This requirement is trivial considering an IC can have billions of transistors.

Moreover, the logic circuits are so fast that your adder can execute billions of addition operations per second.

Now for the other basic operations, namely, subtraction, stretch your imagination to consider $A - B = A + (- B)$. That indicates we could use the full adder for subtraction provided we know how to negate a number. Simply put, we need to include the representation for negative numbers as well.

6.5   Representing negative numbers

Shifting our focus to the world of whole numbers, it is understood that there must be a simple mechanism to represent negative numbers as well; otherwise, many computations involving subtraction cannot be done. We have three different options in the binary domain for an n-bit number.

- *Sign-Magnitude*: MSB (b_{n-1}) is reserved to indicate sign (0 = +ve and 1 = -ve). The rest $b_{n-2} \ldots b_1 b_0$ is the magnitude. The MSB has no weight.
- *1's complement*: same as sign-magnitude but the MSB has a weight of $-(2^{n-1} - 1)$.
- *2's complement*: same as 1's complement but the weight of MSB is -2^{n-1}.

Sign-magnitude, 1's and 2's complement representations

The following Table shows different aspects of the various representations of a 4-bit binary number.

Representation	Weight of MSB	Range	Disadvantages
Normal	$8 \, (2^{4-1})$	0 to 15	No negative number in the range
Sign-magnitude	None	−7 to +7	Multiple representation of Zero (0000 and 1000)
1's Complement	$-7 \, (-(2^{4-1} - 1))$	−7 to +7	Multiple representation of Zero (0000 and 1000)
2's Complement	$-8 \, (-2^{4-1})$	−8 to +7	As such none, used in all computers to represent integers

The following table shows arithmetic operation using 4-bit 2's complement representation of numbers.

---

[1]Your mobile is capable of adding two 64-bit numbers

Operation	Operands	Partial result	Final result	Rules
2 + 3	0010 + 0011	NA	0101 (= 5)	Normal addition
7 + (−5)	0111 + 1011	10010	0010 (= 2)	MSB (end-around is carry or EOC) = 1; Result is +ve. Ignore MSB
5 + (−7)	0101 + 1001	1110	1110 (= −2)	EOC = 0. Result is -ve
5 + (−5)	0101 + 1011	10000	0000 (= 0)	EOC = 1; ignore it
5 + 7	0101 + 0111	1110	1110 (= −2)	EOC = 0; wrong result; +ve overflow
−5 − 7	1011 + 1001	10100	0100 (= 4)	EOC = 1; Ignore; wrong result; -ve overflow

The last two rows of the table show overflow and that is obvious as the results are out of range. To get correct results the operands need to be represented using 5-bits, which has a range of −16 to +15. The programmer must be aware of this deficiency that a fixed number of bits can only be taken into account for any arithmetic operation.

Technically all three representations can be used in computers however hardware and other issues go in favour of 2's complement representation of integers and it is widely used in all sorts of computing devices.

6.6 A full-subtractor

Our next target is to carry out subtraction and if possible using the FA with some extra logic.

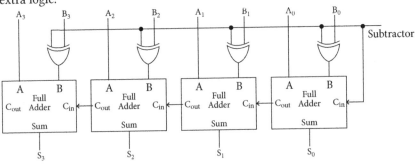

Fig. 6.11: Full-adder cum Full-subtractor. C_{in} = 1 at first stage converts B to 2's complement and we get subtraction A + (− B); for C_{in}=0, B is not converted and we get A + B.
Bit inversion is 1's complement and adding 1 essentially gives you 2's complement.

The subtraction operation represented by A − B; may be considered to be A + (− B). We could convert operand B into its negative form (usually 2's complement form). Fig. 6.11 shows the use of a full-adder with an arrangement to negate an operand to implement a full-subtractor.

6.7 Multiplication and division

Our black box now is capable of doing addition and subtraction — multiplication and division are yet to be included.

Multiplication is nothing other than repetitive addition and so is division where we do repetitive subtraction. Assuming that we have some mechanism to control this repetition then we could carry out multiplication as well as division. This is presented through simple repetition for A × B and A/B, e.g. 11 × 3 and 11/3.

```
S = 0;                           Q = 0;
repeat B times                   repeat until (A-B <= 0)
  S = S + A;                       Q++; A = A - B;
                                   R = A;
/* Note: S = S + A; is algebraically wrong; in computers it means read the
current value of S from memory; add A with it and the result be written back
to memory updating the old value of S. Q++ is the short form of
Q = Q + 1; */
```

Accepting that it would be possible to control repetition by using extra logic circuits we see that our computing black box can carry out all basic arithmetic operations.

In fact, it becomes a functional unit of the Central Processing unit (CPU) and known as ALU (*arithmetic* and *logic unit*) as it can carry out, if required, logic operations as well[2].

6.8 Fractional number and floating point representation

Well, we know that so long as the operations are addition, subtraction and multiplication of integers the result is also an integer. However, we cannot rule out real numbers; the numbers with fractional part that come naturally with division. Chapter 3 deals with real numbers, their properties and many more. So, let us take a look at how a real number can be represented in a computer. Unlike integers, we will not present the hardware used for all arithmetic operations on real numbers as this is not within the purview of this book. Let us stick to get the basics.

6.8.1 Fractional representation

Like the decimal representation any real number with fractional part can also be represented in binary using the negative powers of the radix (= 2) from the right of the binary point (just like after the decimal point for decimal systems). For example $601.75 = 6 \times 100 + 0 \times 10 + 1 \times 1 + 7 \times 10^{-1} + 5 \times 10^{-2}$.

---

[2]In reality special purpose circuits are used for multiplication and division in the CPU

In binary the same is

$$1001011001.11 = 1 \times 512 + 1 \times 64 + 1 \times 16 + 1 \times 8 + 1 \times 1 + 1 \times 2^{-1} + 1 \times 2^{-2}$$
$$= 601_{10} + 0.5 + 0.25$$

So, any fractional number can either be represented exactly or approximated. However, the usual representation of a fractional number as

$$b_{n-1} b_{n-2} \ldots b_1 b_0 \cdot b_{-1} b_{-2} \ldots b_{-m}$$

may not always serve our purpose because this form has the following issues.
- It is not suitable to represent a very big or a very small number.
  - Consider the number 5×2^{62} and it would require the string 101 followed by 62 zeros. Clearly, even a 64-bit representation would fail to represent the number.
- It has accuracy issues as well.

For these reasons, in practice, a special format (IEEE 754 standard) is used to represent any real number and is known to be a *floating point representation*.

6.8.2 IEEE 754 standard for floating point

Though the typical notation of representing a big or a small number was known to us for long (the floating point notation; e.g. Avogadro's Number = 6.022×10^{23} or the charge of an electron = -1.602×10^{-19} C) the same was not formally used in computers until the widespread acceptance of the IEEE 754 floating point standard.

6.8.3 IEEE 754 notation

Typical floating point representation uses 32 bits (single precision) and 64 bits (double precision).

Accuracy	Sign bit (s)	Exponent (e)	Fraction (f)
Single Precision	1-bit (b_{31})	8-bits (b_{30-23})	23-bits (b_{22-0})
Double Precision	1-bit (b_{63})	11-bits (b_{62-52})	52-bits (b_{51-0})

The numeric value of the floating point number is computed as $V = (-1)^s M \times 2^E$, where s is the sign bit; E is the biased exponent and M is the significand.
- A bias (b) $2^{(\text{no. of bits in } e - 1)} - 1$ (127 for single precision) is subtracted from the e-bits to form E (i.e. $E = e - b$).
  [*Note*: if $e = 0$ then $E = 1 - b$]
- The fractional part (f) is stored as a 23-bit string from the implied binary point to the left of the leftmost bit (MSbit) of f. For processing a significand (M) is computed as $M = 1.0 + f$ to form a normalised significand such that $1 \leq M < 2$. This virtual extra bit is used to expand the significand by 1 bit; free of cost.
  [*Note*: M is taken as $0.0 + f$ when e = 0]

The floating point format is ideal to represent real numbers for a wide range with good accuracy. Interested readers may consult Appendix F for more on floating point representation.

6.9 Character and other codes

Besides the numbers, integer and real, computers need to store characters like your name or a piece of poetry or an email. Keep in mind, whatever you store in a computer is always 1s and 0s in the form of binary bit-strings. Some points on character representation follow:

- Most of the tasks managed by the modern digital computer are non-numeric or text (i.e. character level) processing.
- Characters must have suitable representation in the form of numbers (you can only store numbers in the memory).
- American standard code for information interchange (ASCII) is a favoured choice for character encoding.

6.9.1 ASCII code

ASCII was originally designed as a 7 bit code that represented a total of 128 different combinations. When ASCII was designed, computers were in their infancy and the standard input device was a keyboard; known as *teletypewriter* (in short TTY). The objective was to encode small and capital letters 52 (26 + 26) symbols, 10 symbols (0 to 9) and around 20 special characters (punctuations and mathematical symbols, etc.). So, 7-bit code encoding 128 different keys was enough. Now in all computers 8-bit taken together, known as a *byte*, is the minimum addressable unit. This means in a general-purpose computer even if you need a single bit to process you need to read it from the memory as a byte (containing that bit) — process it and write back the byte to memory.

- 8-bit ASCII code is typically used now and the extra 128 combinations are used as control and special characters.
- The mapping of the symbols (say q, T or 1) to a number cannot be random; at least it is to be guaranteed that if the code for A is x then the code for B must be $x + 1$. Similar argument holds good for small a to z and the digits 0 to 9. The space is also a character for a computer and *ASCII* 32_{10} is used to represent it.
- Non-printable characters are to be encoded as well. Thus, carriage return, CR = 13 and line feed, LF = 10. CR is used to move the print-head to the beginning of the line and LF move the print-head to the next line. Together the effect is from a particular print position on a line (say, L) you can move to the beginning (i.e. print-position 1) of the next line (i.e. L + 1).
- There are other special encodings and mostly used for control and other purposes. For example, ASCII 7 produces a bell sound.
  We have provided the ASCII encoding as a list in Appendix F.

6.9.2 Other codes

There are numerous other codes that are in use to serve a variety of applications. In each case some special property of the code is exploited for a typical solution. For example, Grey code where consecutive encodings have difference in only one bit position and have applications that includes encoding of mechanical movement.

6.9.3 UNICODE

ASCII was designed considering only the Roman scripts. Now with the requirement to include virtually all scripts and a variety of graphics symbols (other than the usual characters) a new standard (UNICODE) has emerged which uses 16 bit encoding for any character.

6.10 Organisation of a typical PC

The organisation of the different functional units of a computer is shown in Fig. 6.12. As most people would use PCs we have shown the units typically present in a PC. However, the basic structure is almost same even for a very powerful machine except the fact they use multiple CPUs, more memories and many special purpose units to make the machine more powerful and fast.

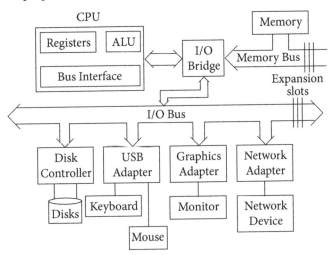

Fig. 6.12: Functional units of a typical PC/laptop along with the peripheral devices

The primary resource of the CPU is memory which is connected through the memory bus. Other than some simple arithmetic and logic operations, the CPU reads from and writes to the memory using the bus. Different types of I/O devices are primarily used for the input and output or we may say the man-machine interface and are all connected to the CPU through I/O bus.

PC is used everywhere by everybody. We believe that the reader may not have any difficulty in understanding why common I/O devices like Monitors are connected through the graphics adapter and simpler I/O devices like keyboard and mouse are plugged to the USB (Universal serial Bus) port (adapter).

With this essential background information now we are ready to learn a programming language to solve arithmetic problems including many problems presented in Chapter 4.

6.11   C programming language: A tutorial Introduction

The objective of this section is to explain how the mathematical or computational logic can be transformed into a computer program with the help of a programming language. Through the program, which is a mechanism to instruct the computer, we describe each step to be carried out in a very precise manner; we call this process *programming* or *computer programming*. Here, in this section you will learn basic programming techniques through small and simple examples.

There are many programming languages used in computers and out of which we have chosen the C[3] programming language. Though C is a small language, it is very powerful and easy to learn. So, we have used C language to solve the problems given in this book. Many other programming languages support "Object Oriented programming (OOP)" concepts; like C++, JAVA, etc. However, for solving typical discrete arithmetic problems there is a little need for OOP techniques.

6.11.1   C programming

Here is your first computer program in C language (source file name is hello.c) that displays the text **hello, world** on the screen.

```
#include <stdio.h> /* C pre-processor directive to include
                    stdio.h header file*/
int main() /* A C-program is a collection of functions.
            One of which must be the main() function */
{ /* Matching curly braces shows a block of instructions */
  printf ("Hello, world\n"); /* Displays the string "hello, world" */
                   /* Newline ('\n') is used to move the cursor */
                   to the beginning of the next line */
       return 0;          /* main() return 0 to its caller */
}
```

Function main() is calling a system supplied library function printf() to display the greetings Hello, world on the monitor. The *return* statement returns 0 to the caller of the function main() which is called by one of the OS (see Section 6.12) supplied functions that initiates the start of the execution of your program.

---

[3]C language was developed in 70's at AT & T's Bell Laboratory — The UNIX operating system is written mostly using C. UNIX/LINUX is considered the most successful and technically competent OS.

The program that you write and keep as a disk-file (usually) is known as the *source-file* (like hello.c). Remember the source file contains nothing but text and consists of a number of ASCII characters (see Appendix F).

6.11.2 Program development cycle and running your program

First we write the program using the keyboard with the help of an *editor* program; i.e. creating the source text file in a particular language (say C or any other higher level language or HLL in short). Then it is translated (conversion from higher level to machine level) by a *compiler* to create a linkable object binary file. For correcting errors, reported by the compiler, we go back to edit the source file. If there is no error, we link the object file with other object files from various libraries using a *linker*. Here also, we may have some problems and we have to go back to the editor or use some advanced tools like debugger for finding out bugs (errors). The linker may successfully create the executable object file and that is loaded from the disk to the main memory by the *loader* program for execution. This process is known as *program development cycle*.

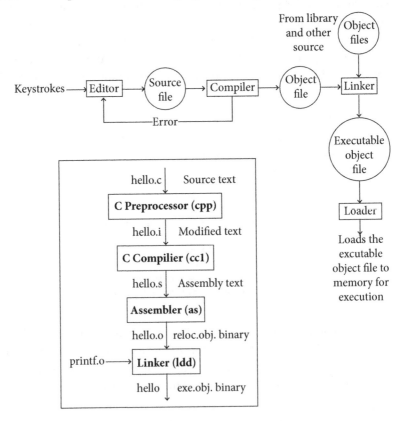

Fig. 6.13: Program development cycle and multi-step translation including linking

In UNIX system the source file (hello.c) may be compiled and run by invoking *gcc* program (known to be *compilation suite*) to create the executable binary file; namely, hello. This is usually followed by loading the same from disk to memory for executing the program by the loader when we type the second command as shown below.

```
unix> gcc -o hello hello.c
unix> ./hello
```

Translation (compilation in computer parlance) from HLL to machine level is a multistep process; usually four including (i) *preprocessing*, (ii) *translation from HLL to an intermediate form known as assembly level*, (iii) conversion of assembly level to machine level by the assembler and (iv) followed by *linking*. The program development cycle and the muti-step translation are shown in Fig. 6.13.

6.12   The role of the *operating system* (OS)

To help execute your program and numerous other managerial tasks, software (collection of many important programs/tools) known as the *Operating System* (OS) is always running in a computer as a supervisor. One of the objectives of the OS is allowing you to develop and execute your program easily through simple commands or some graphical user interfaces. The OS being the manager of the whole computing system is also responsible for proper sharing of the computing resources like, CPU, memory and I/O devices between the competing programs. These programs in execution are known as *processes*. The OS follows a fair and unbiased policy for all the competing processes running concurrently in a computer and manages these processes for an optimized performance.

6.13   A few simple programs

In this section we will see a few C programming examples covering different aspects of the language.

6.13.1   Arithmetic operations

```
int main() /* Addition and division etc., */
{
  int a, b, c, r;
  float f;
  a = 10; b = 3; /* To read the values of a and b
                    we may use scanf("%d %d", &a, &b); */
  c = a+b;
  printf("%d\n", c); /* We may also use  printf("%d\n", a+b); */
                     /* Subtraction and multiplication can be
                        done replacing '+' by '-' and '*' */
  printf("%d %d\n", a/b, a%b); /* integer division; quotient
                        is 3 remainder is 1 */
```

```
    printf("%d %d\n", c, r); /* Print quotient and remainder */
    f = (float) a/b; /* Type casting, (float),  is used
                        to get real division with the result
                        3.333 otherwise the result is as
                        'a' and 'b' are integers */
    printf("%5.3f\n", f); /* Note the change in format */
    return 0;
}
```

Note that to get the fractional part when you divide would be available if you use the proper data type; i.e. float in this case.

6.13.2 Temperature conversion

This program converts a range of temperatures (numbers to be precise) form one scale to another. This also shows how a number of instructions can be repeatedly (a loop) performed.

```
void main() /* Print Celsius-Fahrenheit table */
{
    int min, max, step; /* Integer variables min, max and step are
                           declared here */
    float celsius, fahrenheit; /* celsius and fahrenheit are real numbers
                                  i.e., they can take fractional values */
    min = -100; max = 100; step = 5;
    celsius = min;

    while (celsius <= max) { /* begin while */
        fahrenheit = 9.0 * celsius/5.0 + 32.0;
    printf("%4.0f %6.1f\n", celsius, fahrenheit);
        celsius  = celsius + step; /* This instruction reads the value of
                                      celsius and adds 5 (step) to get the
                                      next celsius value */
    } /* end while */
}
```

The *while* loop continues until *celsius*, which is being increased by *step*, reaches *max*. The conditional expression (*celsius* ≤ max) would be true so long the condition is satisfied; i.e. *celsius* not greater than *max*. The *printf*() function is using some formatting string indicating how we would like to print the values of *celsius* and *fahrenheit*, how many digits in the integer part and how many after the decimal point.

Now let us check the same program using different way of looping. In the previous program, a number of instructions within the body of the while (i.e. between the curly braces {}), is repeatedly executed. Same can be achieved as below with a *for* loop.

```
for (celsius = min; celsius <= max; celsius = celsius+step) {
    fahrenheit = 9.0 * celsius/5.0 + 32.0;
```

```
    printf("%4.0f %6.1f\n", celsius, fahrenheit);
}
```

Here, *celsius* is initialised to *min* only once. The condition *celsius* ≤ max is tested each time the loop is entered. At the end of the loop each time *celsius* is increased by *step*. The control passes to the next instruction after the *for* loop when the condition (i.e. *celsius* ≤ max) becomes false.

6.13.3 Copy input (keyboard) to output (monitor) program

Here is the program to copy whatever you type on the keyboard to the monitor.

```
#define EOF -1
void main()
{ int ch;
  ch = getchar();
  while (ch != EOF) { /* Alternative: while((ch = gethchar()) != EOF)
                          putchar(ch); */
    putchar(ch);
    ch = getchar();
  }
}
```

getchar() and *putchar()* are two matching functions to read or write a character from the standard input and standard output devices (keyboard and the monitor), respectively.

EOF stands for the end of file condition and the operating system returns −1 to indicate the same. To accommodate the negative value of EOF, the variable *ch* is defined as an integer. Note that any character is automatically promoted to an integer in an expression.

6.13.4 Using your own function — other than main()

We will show how a user develops modular programs instead of a big mono-lithic main() only function by having several small functions to be called from the main() as and when required. As an example we will modify the copy input program that would modify all alphabets from small to capital.

```
#define EOF -1
 int main()
 {int c;
   int convert2capital(int ); /* Function template */
   while(( c=gethchar()) != EOF)
   putchar(convert2capital(c)); /* Note conversion to capital is done and
                                    then it is used as the parameter to
                                    putchar() */
      return 0;
 }

int convert2capital(int c)
```

```
{
  if (c >= 'a' && c <= 'z') /* If the character read is small then */
    c = c -'a' + 'A'; /* convert it to capital; see the trick c -'a'
                         computes the offset; e.g., if c = 'b' then
                         c - 'a' = 1, when it is added with the code
                         for 'A' we get the code for 'B' */
  return c;
}
```

In C the parameter(s) passed to a function, e.g. 'c' in convert2capital, is always call-by-value. This means the called *function gets* the value of the parameter passed. It can change it locally but without knowing the address any change is not reflected back to the calling function where the variable (parameter is declared). For this reason if you print the value of the variable 'c' just before the return statement you would not get any change; though for an input of small 'a' (ASCII value 97) it is changed to capital 'A'; (ASCII value 65) in the called *function.*

6.14   Recursive function

Normally, we iterate using a loop construct through a number of instructions many times to solve problems. In many cases, however, the problem statement can be very easily mapped in a recursive way. A classic example is the computation of the factorial of a number. Factorial of n is equal to $n(n-1)(n-2)\cdots\times 2\times 1$. So, we can use the following function.

```
int factorial (int n) /* Iterative version */
{
  int i, f=1;
  for(i=1; i<= n; i++) /* if n = 0; for loop is not done;
                          function returns 1 */
    f = f * i;
  return f;
}
```

The factorial can be recursively defined as $n! = n \times (n-1)!$ with $0! = 1$. This may be implemented by a recursive function (a function that calls itself) as shown below.

```
int factorial (int n) /* Recursive version */
{
  int i;
  if (n == 0)
    return 1;
  else
    return n * factorial(n-1); /* for example say, n = 4 then it is
                                  recursively making it 4 x 3 x 2 x 1
                                  through calling itself each time with
                                  the next  value of the parameter */
}
```

6.15   More programming examples: Counting characters, words and lines

The copy input program can be enhanced to count characters and similar counting programs as it is reading all its input character by character.

6.15.1   Counting characters

```
#define EOF -1
void main() /* Program to count characters */
{ long noc = 0; int c; /* Usually in a long variable more storage
                         is allocated so that the maximum value
                         that can be stored is at least $2^31 - 1$ */
  while ( ( c = getchar() ) != EOF)
  noc++;
  printf("%d\n", noc);
}
```

Here, *while* statement may be replaced by

```
for(; (c = getchar()) != EOF; noc++);
```

Or, we could shorten it even more by dropping the integer variable c as it has no real use here to count characters; so the alternative could be

```
#define EOF -1
void main() /* Program to count characters */
 { long noc;
   for ( noc =0;  getchar() != EOF; noc++);
   printf("%d\n", noc); /* Keep in mind that making it short should not be
                          the goal of a programmer; keeping it right is */
 }
```

Input lines are terminated by '\n' . So we count lines by

```
{ int c, nol=0;
  while ( (c = getchar()) != EOF)
      if ( c == 'n')
            nol++;
       printf ("%d\n", nol);
    }
}
```

6.15.2   Word counting

Word counting can be done in this manner. Assume the white spaces are ' ', \n and \t, i.e. the blank, tab and newline.

A program for counting words

```
#define YES 1
#define NO 0
```

```
void main()
{ int c, now =0; in_a_word = NO;
  while( ( c = getchar()) != EOF) {
    if ( c == ' ' || c == '\n' || c == '\t' )
        in_a_word = NO;
    else if ( in_a_word == NO) {
        in_a_word = YES; now++;
  }
  printf ("%d\n", now);
}
```

6.15.3   Counting alpha-numerals

Consider the problem of counting alpha-numerals in the input. We need 26 + 26 + 10 counters for the capital characters, small characters and numerals. Here is a solution with arrays:

```
...
int small[26], capital[26], numr[10];
for (i=0; i<26; i++){ small[i] =0; capital[i]=0};
for(i=0; i<10; i++) numr[i] =0;
while((c = getchar()) != EOF){
    if ( c >='a' && c <= 'z') small[c-'a']++;
    else if (c>='A' && c <= 'Z') capital[c-'A']++;
    else (c>=0 && c<=9) numr[c-'0']++;
}
```

The Array type: Here in this program we have introduced a new data type, namely, the array, which is useful in many cases. An array is defined using the base type and expressing the number of such items to be held by the array. Thus, *int student*[100] is a declaration that the array variable *student* will have 100 integer elements, student[0], student[1] to student[99] and in general any element can be referred to as student[i]; *i* will always be an integer serving as an index to the required element. In C the array index always starts with 0. Indeed, arrays of other types and multi-dimensional arrays are also possible.

In the program above first we declare 3 arrays for all 62 characters. These arrays are effectively categorising the three types; capital, small and numerals. Moreover, also alleviated is the problem of using 62 different variables and an easy access to any one of the elements with the array_name[index] format. Then each counter is set to 0 using two for loops. A single *while* loop reads all characters at the input and *If-else* is classifying the input. ASCII code of 'a' is 41H, 'b' is 42H, 'A' is 61H, 'B' is 62H. Similarly, '0' is 30H while '1' is 31H.

6.16   Data objects, constants and declarations

Variables and constants are the basic data objects manipulated in a program. Declarations list the variables to be used, their types and their initial values. Operators specify what is to be done with them.

Variable names: Names made up of letters and digits; the first character should be a letter. Underscore "_" is a valid character and is used to increase the readability. Upper and lower-case letters are different. Traditional C practice is to use lower case for variables and all upper case for symbolic constants.

C Provides a number of basic data types; they are:

- *Integer*: It has 4-forms *int, short, long* and *unsigned*. By definition short is guaranteed to be allocated same or less no. of bytes than that of the integer. And long would get same or more space in memory than *int*. In many implementations *short* gets 2 bytes, *int* gets 4 bytes and *long* gets 8 byte space.
- *Character*: *char* stores everything in a byte.
- *Real*: *float* (single precision: 4-bytes) and *double* (8-bytes)

The construct 'x' indicates a character constant and is stored as a single byte ASCII value; i.e. the ASCII value of x. Certain non-graphics characters represented by the escape sequence like '\n' (newline), '\t' (tab), '\0' (null), '\\' (backslash) and '\'' (quote). An arbitrary byte size bit pattern can be used as '\ddd', e.g.

```
#define FF '\014' /* ASCII form feed -- used to feed a new page */
```

The construct "x" on the other hand is a string (multiple characters together) constant which occupies two bytes in memory; i.e. code for x followed by the null (\0) character which is the end of string character. Thus, "ABC" is a string constant of length 3 but takes 4 bytes in memory.

Symbolic constants may be defined as:

```
#define THISYEAR  2021
#define PI 3.14
```

A real constant can be written in scientific notation as: 123.456e7 or 0.123E3 Floating point constants are always promoted to double. Long constants are written as 123L. Large *int* constants are promoted to long. Octal constants are written as 037 (octal) or in hex as 0x1F (or 0X1F); 037L or 0x1FL to make them long.

All variables must be declared before they are used.

```
int lower, upper, step, i=0;
char c;
float epsilon = 1.0e-10;
int age; char backslash = '\\';
```

Note that a declaration specifies a type, and is followed by one or more variables. Initialisation, if required, can be made with the declaration.

6.17   Arithmetic, relational and logical operators

We have already seen the standard arithmetic operators like, +, - etc. The *pre-* and *post-* increment and decrement operations may be elaborated.

++ and -- are handy forms of incrementing or decrementing variables. They can be used both as prefix or postfix. Take care to note the difference: Say *n* = 5.

- x = ++n; would assign 6 to x (i.e. increment first and then assign); and
- x = n++; would assign 5 to x (i.e. assign first and then increment).

In both the cases 6 will be assigned to n.

So, for simple increment use of postfix or prefix form does not matter. However, there are situations when one form or other is specifically used.

There are six relational operators; >, >=, <, <= (same precedence); just below them are the equality operators == and ! = . Precedence stands for the order of association and execution of the arithmetic or logical operations in an expression.

Logical operators are && (AND), || (OR), and ! (NOT). Expressions connected by && or || are evaluated left to right and the evaluation stops as soon as the truth or falsehood of the result is known.

6.18   Bitwise Operation

In C any value other than 0 is considered TRUE; so if a = 5 and b = 2, then (a && b) yields TRUE. On the other hand (a & b) would yield 0 as the result. Note the internal representation of 5 and 2 are 101 and 010 respectively. So, bitwise ANDing (&) would set all these 3 bits to 0.

Bitwise logical operators are (i) ˜(not), (ii) & (and), (iii) |(or), (iv) ^(xor), (v) <<(left-shift), and (vi) >>(right-shift).

The operator >> shifts the operand by one bit position to the right and the vacant space is filled with 0. So, the operator >> may be used for integer division by 2 and << may be used to accomplish integer multiplication by 2.

These operators are not applicable to *float* or *double*. To illustrate the use of bit operators let us examine the following functions mul4() and div2().

```
int mul4(int n)
{
 return ( x << 2); /* returns n x 2 as the left shift is
                      essentially multiply by 2 and faster
                      than using multiplication */
}

int div2(int n)
{
 return ( x >> 1); /* returns the integer division result
                      as the right shift by 1 is essentially
                      divide by 2 */
}
```

6.19   C Keywords

There are some keywords in C which cannot be used as identifiers (a variable or constant). For example, *int, for, while,* etc. may not be used as identifiers because they have got predefined meanings and functions in a C program. See Appendix F for the list of C keywords.

6.20   Control flow

By now, it is understood that in a program the flow of execution is controlled by the type of statements used. In normal cases, this flow is typically sequential. For all practical purposes this sequential flow need to be broken. And we may either loop, i.e. repeatedly execute a group of instructions on some condition; or branch to skip the next statement(s); usually on a condition to skip a block of statements.

We have already seen *while* and *for* loops and there is a third version known as *do–while*.

6.20.1   Loop statement: do ... while

The *while* statement checks the condition at the beginning; if it is FALSE the body of the loop may not be executed at all. In some cases it is needed that the body of the loop should be entered at least once. Such a need can be catered through *do–while* loop statement as shown below.

```
void main()
{ int a, r;
   a = 100;
   do {
        r = a % 2;
        printf("%d", r);
        a = a / 2;
        }
      while ( a > 0);
}
/* Note that this program converts a decimal number to binary in
   reverse order */
```

In order to make an exit from a loop we may use break, e.g.

```
for (i=0; i < 100; i++){
    <statement>
    :
    if ( <exp>)
            break;

    <statement>
}
```

This loop body which would otherwise be executed 100 times breaks if the condition is true (break is executed).

continue is used to bypass a group of statements in a loop, e.g.

```
for (i=0; i < 100; i++){
    <block1>;
    if (<exp>)
    continue;
    <block2>;
}
```

Here, block2 will be bypassed and loop will be started from the beginning with the next value of *i* if the condition gets satisfied; i.e. continue is executed.

6.20.2 Branching

In programs the statements are executed one after another in a sequential manner. Within the body of the loop the execution of the statements is also sequential. However, depending on certain events we may need to break this sequential nature, and we may not execute the next instruction in the sequence and start from elsewhere. This is done with the decision statement; namely, *if* statement. The *if* statement may or may not have the *else* part. Both the cases are shown below.

```
c = 20;                                     c = 20;
if ( a > b)  /* depending on the condition we     if (a > b)
    c = 10; /* change the value of c to 10 */          c = 10;
                                            else
                                                c = 5;
```

The bifurcation of execution path shown in the second case above has a shorter form if we use the 'ternary operator' (c ? (a > b) = 10 : 5;). Here, if the condition is true then *c* gets the value 10 else 5.

Multiway branch is achieved by using switch statement: e.g.

```
while (( c = getchar()) != EOF){
    switch c {
        case 0:
        case 1: <block1>;
            break;
        case 2: <block2>;
            break;
        default: <block3>;
            break;
    }
}
```

Suppose the input *c* is the character 0 or 1, we would like to execute instructions within block1. If *c* is 2 we would like to execute the instructions within block2. For all other values of *c* we would take the default action, i.e. execute the instruction inside block3. Note that the last break is not mandatory.

The instruction *goto* is used for an unconditional branch. Unconditional branch from one point to another is an undesirable feature of any modular program and it is generally avoided. However, it may be the only way-out if you need to come out of a deeply nested block, e.g.

```
{<b>;
{<b>; {<b>; {goto getOut;}
 <b>; }<b>;}<b>;
};
getOut: <b>;  /* getOut is considered a symbolic label */
```

In this section we did not provide formal discussions or working examples on many aspects of the C language; this includes (i) local and global variable; (ii) the scope of a variable; (iii) storage class; (iv) structures; and (v) pointers stated. The reason is that by and large they may not be required to solve problems. And in a few cases the reader can pick up the required knowledge from the solutions to the problems (Chapter 7) indirectly. However, considering the importance let us see what are the pointer and local variable.

6.21   The pointer type

C has a special data type known as the *pointer*. By definition a pointer is a variable that holds the address of another variable. Whenever you refer to a variable, it may be assumed that the machine generates the address of the variable to read or write. So, if an expression is A = B + C; addresses of the variables B and C are used to read them and the result of addition is written to another variable (A here) using its address once again. Now this type of direct addressing may not be a preferred choice for a variety of reasons. Here comes the use of the pointers. Let us examine them in a proper perspective. See the following program.

```
int *pa, a, b; /* pa is a pointer variable and would be pointing
                to an integer; a, and b are normal integer variables */

a = 10; /* suppose a is stored in memory location 100 */
pa = &a; /* The unary operator & returns the address of the
            variable a (in this case 100) */
b = a; /* b gets the value of a */
b = *pa; /* b gets the value of a through an indirection; the pointer
            technically the last two instructions have the same effect */
```

6.22   Local variables

Whenever you define a variable in the body of a particular function, it becomes local to that function. The local variables (unlike global; those are defined outside the body of a function), come and go with the invocation of the function, and they do not exist when you make an exit from the function. Let us see this point through an attempt to swap the value of two integer variables in a program.

```c
int main()
{
  int a,  b, temp;

  a = 10;
  b = 5;

/* to swap or interchange we may use the following code */
  temp = a;  /* a is stored in a temp */
  a = b; /* a gets the value of b */
  b = temp; /* b gets the old value of a set aside in
                temp for swapping; so a is 5 and b is 10 now */
 return 0;
}
```

Let us try to write a function to do the same job;

```c
void swap(int a, int b)  /* wrong version */
{
  int temp;

  temp = a;  /* a is stored in a temp */
  a = b; /* a gets the value of b */
  b = temp;
```

We try to swap using the newly written swap function and see that it is wrong.

```c
int main()
{
    a = 10;
    b = 5;
    swap(a, b);
    printf("%d %d\n", a, b); /* we will get 'a' is 10 & 'b' is
    5 and no swapping */
    ...
}
```

The reason is that in C the called *function* gets the value of the parameters (a and b here) and not their address. Hence in the swap function the interchange is done locally but it has got no permanent effect; because the addresses of a and b are not known (or *passed*) to swap. Here is a solution using pointers.

```c
void swap(int *pa, int *pb)
{
  int temp;
  temp = *a;
  *a = *b;
  *b = temp;
```

```
}
```

```
/* now in the main you call swap as shown below */
```

```
    swap(&a, &b); /* note that the addresses are now known to swap and
                    it can change the values using the address */
```

The reader may be warned that though we are passing pointers (addresses) to a function logically the function is still call-by-value as the value of the pointers are passed using which the called *function* can change the value of the variables.

6.23   Type conversion

When different types of operands appear in a single expression they are converted to a common type that make sense, e.g. expression like i + f (i : integer and f : float); i is converted to float. In f + d, f is converted to d (double). For example, *int* and *char* can be freely intermixed.

```
void atoi (char s[]) /* string  to integer */
{   int i, n = 0;
    for (i = 0; s[i] >= '0' && s[i] <= '9'; i++)
      n = n * 10 + s[i] - '0';
    return n;
}
```

6.24   The task ahead

By and large with this rudimentary knowledge of C programming you may start solving simple problems and then try to solve some of the problems of Chapter 4 that would automatically sharpen your programming skill. Moreover, through this you will be dealing with the numbers more intimately getting more insight and fun.

There are many good *e*-books and tutorials available and you may use them to broaden your knowledge on programming in general and C-programming in particular.

Finally, this tutorial is only for the readers to start writing programs. Indeed, many topics are omitted to make it concise. However, the discussions and example programs may be enough to solve most of the problems given in this book.

Chapter 7

Programming Solutions

In this last chapter of the book, we present how a computer can help you solve arithmetic problems using appropriate programs. The readers must have noticed the subtle message that we give more emphasis to analytical solutions to arithmetic problems. However, if you are running short of ideas to produce an analytical solution, a computer may be a handy tool in many cases. We also know that in some cases a computer cannot help in solving the problem in your lifetime. Moreover, we would see that a naive program may not produce the result, or it may take too much time. Improving the performance demands adding analytical logic to our programs to such an extent that it becomes an analytical solution.

We will start with preparing ourselves to write some functions of our own that may prove useful. Here, we start with a bottom-up approach; i.e. try to figure out the functions that would be used often to perform simple arithmetic for solving problems. Moreover, for building any new function we will try using the already developed functions. This kind of modular approach is the key to rapid software development. We will also see how they can be kept in a library as a resource to be used as and when necessary. Finally, you may notice that any kind of error checking is not included in our samples. For example, what happens if you input a negative number to compute a factorial. Professional quality functions should have error checking and reporting options. Here, we have kept everything simple for obvious reasons.

7.1 Some useful functions

Odd or even

Finding out if a number n is odd or even may be useful in many situations. Two different versions, namely the division method and bit-testing method, are presented; obviously the bit-testing version is much faster.

```
int oddOReven(int n) /* Division */ | int oddOReven(int n) /* Bit-testing */
{ return  n % 2; }                   | { return n & 1; }
```

We know that bit 0 (b_0) of the binary representation of an integer n will be 0 if it is even. This is similar to testing the digits in unit position (i.e. b_0) of a decimal number where the possibilities are 0, 2, 4, 6 and 8. In binary it is only 0 that is to be tested instead of the possible 5 in decimal[1]. Moreover, division takes more time than bit testing.

Find factorial

A common function that would be required to solve many problems.

```
int fact(int n) /* This function returns factorial of n.
                    No error checking on input is done */
{
    int i, f = 1;

    for(i=2; i<=n; i++) /* For n = 0 or 1
                            we will not enter the for loop and
                            f = 1 will be returned */
        f = f * i;
    return f;
}
```

A factorial can be recursively defined as $n! = n \times (n-1)!$. Note that $1! = 0! = 1$. This leads easily to another kind of implementation of the function where a function is calling itself and known as a recursive function.

```
int fact(int n)
{
    if (n == 0 || n == 1) /* The mechanism may be understood
                             easily with n = 4 */
        return 1; /* fact(4) = 4 * fact(3) = 4 * 3
                      * fact(2) = 4 * 3 * 2 *fact(1) */
    else          /* = 4 * 3 * 2 * 1 */
        return n * fact(n-1); /* Calling itself again with 1 less than
                                 the last value of n */
}
```

Some mathematical operations can be implemented very easily as recursive functions. However, recursive functions takes more memory resources as the partial results are all temporarily stored in the stack (a portion of the main memory). Moreover, calls and returns take more time than any other instructions. Finally, care must be taken to frame the condition to return in such a manner that after certain steps it naturally stops (Here n is reduced in each step and finally becomes 1 to return) otherwise the function would be calling itself endlessly and the program would crash due to stack overflow.

---

[1] radix 10 versus radix 2

GCD or Greatest Common Divisor

Well, you have done it right from your early school days using the division method. Here is a classic solution from the great Greek mathematician *Euclid*.

```
int gcd(int a, int b)
{
    while (a != b)
      if (a > b) a -= b; /* prefix notation
                            same as the postfix a = a - b;  */
      else
              b -= a; /* You can also use the continued division */
                      /* technique  learned  in the school; shown next */
    return a; /* or b as both are same; i.e., the GCD */
}
int gcd(int a, int b) /* Division method, naturally much faster */
{   int r;
    while ( 1 ) {
        if ( (r = a % b ) != 0) {
            a = b; b = r;
        }
        else
            break;
    }
    return b;
}
```

One may notice that GCD too can be implemented recursively. Here is the function.

```
int gcd(int a, int b)     /* recursive version */
{
      if ( b == 0 )
        return a;
      else
         return gcd(b, a%b);
}
```

While recursive way is more elegant it may not be always easy for the beginners to perceive the idea. So, we have presented all other functions in a non-recursive way.

Finding factors

For any integer n we always have a number of factors. For $n > 1$ you have two or more factors. The numbers with exactly two factors (1 and the number itself) are known as prime otherwise composite.

The function presented here fills-in the array facs[] with the factors of the number n and returns how many factors are present. So, the function getFACTORS(24, fac) returns 8 with fac[0] = 1, fac[1] = 2, fac[2] = 3 ... fac[7] = 24; i.e. that is all possible factors 1, 2, 3, 4, 6, 8, 12 and 24.

```
int getFACTORS(int n, int fac[])
{
    int i, p =0;
    for(i=1; i<=n; i++)
      if ( n % i  == 0)
          fac[p++] = i;
    return p;
}
```

Extracting digits from a number

We present two varieties. (i) Extracting the digit in the unit's place and (ii) Extracting digit from any position.

```
int getunitDigit(int n)
{
   return n % 10;
}
```

The digit extractor function can be generalised to return a digit in specific position from the right:

```
int getDIGIT(int n, int pos)
{
   int i;
   for (i=1; i <pos; i++)
      n = n/10;
   return n % 10; /* could be call to getunitDIGIT(n) as well */
}
```

You may need a function that returns the number of digits in 'n'.

```
int howmanyDIGITS(int n)
{
   int d=0;           /* Alternatively ; int d; */
   while(n!=0){        /* for (d=0; n!=0; n=n/10) */
   d++;               /* d++; */
   n = n/10;
   }
   return d;
}
```

Is prime?

Finding out whether a number *n* is prime or not is frequently required. Note that this can be done by using the factor finding function, getFACTORS(), as for any prime we will have only two factors. There may be a straight-forward approach as well; like dividing the number with divisor like 2, 3, up to $\sqrt{n}$; beyond which you need not go.

```
int isPrime(int n) /* returns 1 if prime 0 otherwise */
{
    int  i, last, p = 1;
    last = (int) sqrt(n); /* This (int) makes the float return value
                             from sqrt() to an integer. Note sqrt(n)
                             returns positive value only */
    for (i=2; i <= last; i++)
      if ( n %  == 0){
        p = 0;
        break;
    }
    return p;
}
```

Find a^b

For computing a^b a function, namely pow(), is available in the built-in math library of any C compilation system. However, we can have our own power() function.

```
int power(int a, int b)
{
    int i, p = 1;
      for(i=1; i<=b; i++)
         p = p * a; /* for b =n; p = a * a * ... *a; n-times */
      return p;
}
```

Prime Factorisation

We have learnt to express any number as a product of prime factors. In fact that is the basis of finding LCM. We apprehend such a function would be useful in the present case as well. Note that we may very easily determine the prime factors of a number using our getFACTORS() function and testing the factors thus obtained by isPRIME() function. This shows a very important way of building a new function using the already developed functions. This modular approach is key to hardware as well as software development.

Let us see the problem to be solved. Take, $115 = 5 \times 23$. So, both the factors are primes and required only once in product of prime factors equalling 115; However, 24 yields 6-factors excluding 1 and 24. The factors are 2, 3, 4, 6, 8, and 12. Except 2 and 3 none of them are primes and the prime factor representation is $24 = 2^3 \times 3$. So, we have to find out both the primes and how many times they appear in the product. As an example, for 24 it would be $2(3) \times 3(1)$. Here is the program which returns this information in the array as pairs (occupying two consecutive cells in an array).

```
#include <stdio.h>
int primeFACTORISATION (int n, int fac[])
```

```
{
     int i, j, k, l, tf[20], tn, p, m;
     int isPRIME(int );
     int getFACTORS(int, int []);
     p = 1;
     tn = n;
     j = 0;
     l = getFACTORS(n, fac);

     for (i=1; i < l-1; i++){  /* didn't take  1 and the number itself */
        if  ( isPRIME(fac[i]) == 1 ) { /* it is a prime factor */
          k = 0;
          while (n % fac[i] == 0){ /* Compute how many times the prime factor
                                      is required to produce the number */
               k++;
               n = n / fac[i];
          }
          tf[j++] = fac[i];    /* store the pair in a local array */
          tf[j++] = k;

          for (m=1; m<=k; m++)  /* Check if you need the next prime
                                   factor or not */
               p = p*fac[i];

          if (p == tn)          /* Prime factoring done; break */
             break;
        }
     }
     for(i=0; i < j; i++)      /* copy the pairs to the arrya fac[] */
          fac[i] = tf[i];

        return j;              /* Return the no. of items copied to fac[] */
}
```

For a call like primeFACTORISATION(24, fac) — the fac array would be $\{1, 2, 3, 4, 6, 8, 12, 24\}$ due to the call to getFACTORS and then it will be rewritten to $\{2, 3, 3, 1, 6, 8, 12, 24\}$. See that the first 4 entries of this array are actually required and this value 4 will be returned to the caller of this function for further processing.

Array of primes

This function fills an array with primes in the range of 2 to any integer n. The algorithm used is simple and elegant and known as *Eratosthene's Sieve* — a tribute to Greek mathematician and philosopher Eratosthenes (c.267 BC c.195 BC). Here we start with an array that is initialised with $2, 3, 4, 5, 6, \ldots, n$. The first non-zero number encountered in this array is a prime (in our case 2). We then erase all the multiples of 2 starting with 4. The next non-zero number after 2; i.e. 3 is declared prime and all its multiples starting with 6 are removed. Note that 6 being 2×3 has already been removed as a multiple of 2, so the first multiple that is really removed this time is 9. This is repeatedly done and a count is kept until all have been removed. The numbers remaining or sieved out from the initialised array are all primes in the range of 2 to *n*. We may have two different ways to keep the primes in the array p[]; i) In a compact form where p[0] = 2, p[1] = 3; ... or ii) we organise the sieved out elements of primes and keep them in their respective indices (0 to maximum integer *n*); such as p[0] = 0; p[1]=0; p[2] = 2; p[3] =3; p[4] = 0; p[5] = 5; p[6]=0; p[7] = 7; p[8] = 0; p[10] = 0 and so on. The second option, though wasting space, is helpful in searching a particular prime through the index; for any number *n* if the element p[n] not equal to 0 it is a prime.

```
int eratosthenesSIEVE(int n, int x[]) /* Sieve Algorithm; after sieving
                                         if x[n] != 0 it is a prime */
{
    int i, j, c, t, f, ft;
    x[0] = x[1] = 0; /* primes start from 2 so */
    for(i=2; i <= n; i++)
      x[i] = i;  /* keeping a number in a slot whose index = number */
    c = 0; /* Array is full */  j = 0; /* prime count */ i = 0;
    while (c < n-1){ /* all ( n - 2 + 1 = n-1) array elements are sieved? */
      while (x[i] == 0) /* skip the entity already sieved */
          i++;
      /* first non-zero entity is a prime */
      j++; c++;
      f = 2;     /* now remove its multiples from the array */
      t = x[i++]; /* skip past the present index containing a prime */
      ft = f * t;
      while (ft <= n) { /* ft is within the permissible index */
        if(x[ft] != 0){ /* remove the multiple; if not already done */
          x[ft] = 0;
          c++;   /* increase the sieve count */
        }
        f++;
        ft = f *  t; /* next multiple */
      }
    }
    return j; /* Prime count */
}
```

In testing the Goldbach's conjecture, discussed later, we will use the sieve algorithm for generation of primes in an array.

Random numbers

In many applications including a number of problems in this book, we need to generate random numbers to solve our problem. In a computer some mathematical technique is used, and the generated random numbers are known as *pseudorandom numbers* or PRN, as the sequence of such random numbers may repeat in different execution of the program. The simplest example of true random number is the throw of a six-face dice generating the number 1 to 6. A C-library function *rand()* generates random number in the range of 0 to 32767. There are many techniques used to generate PRN; one of the oldest methods is linear congruential method that is used to generate PRNs within a particular range.

The recurrence relation used in this case is $X_{i+1} = a \times X_i + c \bmod m$. Where, X_i is the sequence of PRN, $m(> 0)$ is the modulus, $a((0, m))$ is the multiplier, $c((0, m))$ is the increment and $X_0([0, m))$ is the start value of the sequence known as the *seed*. For a particular seed value the sequence of PRN will be the same — a highly undesirable property for random numbers. Thus, a built-in function *srand()* is sometimes used to randomly generate the seed value. In order to ensure that *srand()* generates different seed value in each run the system time may be used as the parameter to ensure separate seed. Here is a program to generate the throw of a dice.

```
#include <stdio.h>
#include <stdlib.h>
#include <time.h>
int main() /* Program displays 5 sets each containing 4 of random numbers
               to mimic 4-players playing a board game */
{
srand(time(NULL)); /* always with a different value */
    for(int i=1; i<=5; i++){
      for(j=1; j<=4; j++)
        printf("%d ", rand() % 6 + 1);
      printf("\n");
   return 0;
}
```

While we recommend using the library functions *rand()* and *srand()* you may also have your own random function; say *myrand()*.

```
#define a (1103515245)
#define c (12345)
#define m (1<<31)
static unsigned int seed = 1;
void srand(unsigned int s) {
seed = s;
}
int myrand() {
seed = (a * seed + c) % m;
return seed;
}
```

At this juncture we see the list of functions to be used is growing. But at the same time we also notice that writing any new function using the modular approach of using already developed function makes our task relatively simple. But we need to keep these functions handy, possibly in a library, to use them in any program. So, sharing of the same functions across many programs is guaranteed. Moreover, just like in a real library, the collection grows over the years.

Your function library may have a modest beginning with a couple of functions. Soon, new functions would be built, tested and added to the library enriching it to such a point that a new problem can be tackled by calling appropriate functions. Creating your own library is presented in Appendix G.

7.2 Solution to simple problems

Let us see the solutions to some simple arithmetic problems. In the first example we do not need any function developed so far.

Representing a fraction as the difference of two fractions

Take $\frac{1}{32}$ as an example. Can we represent it as $\frac{1}{a} - \frac{1}{b}$ where a and b are integers? If we can then how many such solutions are possible. Analysis may produce the solution. However, here a computer program can help.

```
int main()
{
  int i, j;
  for ( i=1; i<1000; i++)
    for(j=i+1; j<1000; j++) /* j must be bigger than i else the difference
                                      would be negative */
    if ( ( (float) 1/i ) - ((float) 1/j) == 1.0/32.0 ) /* As 'i' and 'j'
              are  integers we will be getting integer division result (0 in
              this case) unless we force them to a float by putting (float)
              in front of the expression; this change is known as casting
              Also, note 1/32 is integer division but 1.0/32.0 is float */

         printf("(%d, %d) \n", i, j);
  return 0;
}
```

The output is (16, 32), (24, 96), (28, 224), (30, 480) and (31, 992) when i and j both vary from 1 to 1000. Another important question is why we are considering up to 1000. Note that 1/32 = 0.03125. In that case 1/a or 1/b would be at the most 0.001 if we take $i = j = 1000$. This fact is corroborated by the result where the maximum value of a turns out to be 31. So, going beyond 1000 is not necessary. Try with $i = j = 100000$ or more you will be getting the same set of 5 answers. Technically if you make $a = 32$ or more then there will be no solution for obvious reason and thus the execution time may be reduced by varying i from 1 to 31 only. However, as mentioned earlier, the solution depends on the execution and repetitive action. We may not always try to make our program smart.

How many prime factors are there in 11!

$11! = 11 \times 10 \times 9 \times 8 \times 7 \times 6 \times 5 \times 4 \times 3 \times 2 \times 1$ and the prime factors are 11, 7, 5, 3, and 2. The programming solution is very easy using the already developed functions namely, getFACTORS() and isPRIME().

```
int main()
{
   int getFACTORS(int); /* Template for this getFACTORS() */
   int isPRIME(int);     /* Template for isPRIME() */
   int n;
   int f; /* factor count */
   int pf =0; /* prime factor count */
   int fac[100]; /* space reserve for 100 factors */
   printf("Input a positive integer\n")
   scanf("%d", &n); /* Read the number */
   f = getFACTORS(n, fac); /* f = total no. of factors including
                                  prime factors */
   for (i=1; i<pf; i++)
       if (isPRIME(fac[i]) == 1) /* Is it a prime factor */
           pf++;
   printf("The number of prime factors in %d = %d\n", n pf);
   return 0;
}
```

Fibonacci numbers

Fibonacci numbers have unique properties and a lot of possible applications. Here we provide a solution to generate Fibonacci numbers. We have usually avoided recursive solution but this is an exception. Iterative solution is also provided.

```
#include  <stdio.h>
int main()
{
  int fibo(int );
  int n = 20; /* read n if required */
  for(i=1; i <= n; i++)
    printf("%d ", fibo(i));
  printf("\n");
  return 0;
}
int fibo(int n)                  | int fibo(n)
{                                | {
  if (n == 0 || n == 1)          | int f1=0; int f2=1;
    return n;                    |   if ( n==0 || n == 1))
  else                          |      return n;
    return (fibo(n-1) + fibo(n-2); |   for(i=2; i<=n; i++){
}                                |      f3 = f1 + f2;
                                 |      f1 = f2; f2 = f3
                                 |   }
                                 |   return f3;
                                 | }
```

As explained earlier the recursive solution in this case is compact and elegant. This program generates up to the *n*th Fibonacci number where we have assumed that the first two are 0 and 1.

Sum of digits in 11!

Programming solution is again easy and rather tempting for a bigger n!.

```
int main()
{
    int f,  sum=0;
    int fact(int );
    int getunitDIGIT(int );
    int l=0; /* position */
    f = fact(11);
    printf("The factorial is %d ", f);
    prinf("%d, ", f);
    while ( f ) { /* if f <> 0 the loop continues */
            sum = sum + getunitDIGIT(f);
          f /= 10;
    }
    printf("% and the sum of the digits = %d\n", sum);
    return 0;
}
output: The factorial is 39916800 and the sum of the digits = 36
```

The solutions given in this section mostly depend upon using the functions developed earlier. We hope that this section would also help build your confidence in approaching a solution to a problem in a modular manner. Next, we discuss the advantage of using a computer for a repetitive job.

7.3   Computers are good at repetition

Computers are very fast and are not bothered, unlike humans, by repetitively doing the same job. Here is an example in terms of computing the value of π. It has long been known that the ratio $\frac{355}{113}$ is a good approximation to the constant π. Now using a more common (but less correct) ratio $\frac{22}{7}$, we get $\pi = 3.1428571428571\ldots$. The recurring part (142857) is repeating after every 6th digits. You may easily understand that dividing something by 7 generates a remainder from 0 to 6. As 22 is not exactly divisible by 7 the remainder cannot be 0. So, with each step we will be getting, say, different remainder out of the one of the possible 6 remainders. But this is bound to repeat on the 7th digit after decimal point as all are exhausted in the last 6 divisions. And hence, repetition begins.

```
7)22(3.142857
  21
  ──
   10                       1st remainder after multiplying it by 10
    7
   ──
   30
   28
    ──
    20
    14
    ──
    60
    56
     ──
     40
     35
      ──
      50
      49
       ──
       10                   Repetition so the sequence 142857 will appear
```

One might be interested to explore the same for $\frac{355}{113}$; if you are not so lucky, the repetition may start on 113th division. Let us present a program and its output to see the result.

```c
#include <stdio.h>
#include <stdlib.h>
int main()
{
    int p, q, r, i,j, n, d, qwhole;
    int rems[113]; /* keeping remainders here */
    n = 355; d = 113;
    qwhole = n / d;
      printf("The whole part is %d.\n\n", qwhole); /* Printing the whole part */
    for(j=1; j<= 2; j++){ /* Catch the repetition 2 times */

  for(i=0; i<113; i++)
     rems[i] = 0; /* Remainder array initialisation */

      printf("The fraction part \n"); /* Printing the fractional part */
      /* Anticipating a recurring decimal */
      p=0; /* repeat count */
      for(i=1; i< 113; i++){
         /* working with the fractional part only */
         r = n % d;
if (rems[r] == 0)
         rems[r] = r;     /* a new remainder */
else
  break;
while ( r < d )
        r = 10 * r;
n = r;
p++;
q = n / d;
printf("%d", q);
    }
    printf("Repeats  %d places after the decimal point \n\n", p);
    n = 355; d = 113;
```

```
    }
    return 0;
}
```

For this type of problem, computers are really good at dealing with repetition. Don't try a manual division to see when it repeats. So, we may transfer the drudgery of inefficient manual computation using a simple program. We are also amazed to find out that for $\frac{355}{113}$ repetition occurs ONLY after 101 decimal digits in the fraction as shown below.

The whole part is 3.
The fractional part is
141592923539823884955752212389385
3973451327433628318584779646176
991154424778761619469265486725665
37168
Repeats 101 places after the decimal point
141592923539823884955752212389385
3973451327433628318584779646176
991154424778761619469265486725665
37168

The program may be slightly altered to check repetition for other fractions as well.

The time has come to discuss how computers help in checking the validity of conjectures and finding counter examples to prove a conjecture as incorrect. Conjecture demands — to prove or disprove, a very large solution space with infinite number of input variations. So, a blind programming approach is often inadequate to get the verdict within an acceptable amount of time.

7.4   Conjectures

Here we discuss conjectures whose formal proofs still elude us, like Collatz and Goldbach conjectures, conjectures for which we could find a counter example, negating the validity like Euler's conjecture and conjectures which have ultimately been proved, such as, Fermat's last conjecture/theorem.

Collatz Conjecture

Start with a positive integer; if it is even, the next integer will be one-half of the previous one, else (i.e. if it is odd) it will be three times the previous one plus 1; and continue this way to get a sequence of integers. The sequence, irrespective of the starting value of the integer, always ends with a repetitive sequence of 4, 2, and 1; the attractors described earlier.

```
#include <stdio.h>
```

```
int main()
{
    int  n,k0,  k1, k2, k3;
    k1=k2=k3=0;
    printf("Choose a +ve integer\n");
    scanf("%d", &n);
    printf("%d ", n);
    k0 = n;
    while (k1!=4 || k2 !=2 || k3 != 1){
            if (k0 % 2 == 1) /* Here we can use evenORodd() function */
                k0 = 3 * k0 + 1;
            else
                k0 = k0/2;
            k3 = k2;
            k2 = k1;
            k1 = k0;
            printf ("%d ," k0);
    }
    return 0;
}
```

Goldbach conjecture

Here, we will see that the Goldbach conjecture holds good up to a counting number that can be expressed easily in a typical computer. To remind the reader, Goldbach conjecture says that any even counting number greater than 2 can be expressed as the sum of two primes. For example, $4 = 2 + 2$; $6 = 3 + 3$; $10 = 5 + 5 = 7 + 3$; etc. The following program does the checking right from m (≥ 4) to n (A large integer) that a typical computer (PC) supports. The approach used is picking up primes, two at a time, from an array of primes already created through Eratosthenes's Sieve algorithm, and adding them to make a match with the whole number (even) being tested. Note that the search time increases as we deal with bigger numbers. Moreover, there may be multiple solutions (like 10 has two solutions) increasing the computational burden if we care to see all possibilities. The following program shows the result for the range 4 to 100000 and we observe that the conjecture happens to be correct in this range. Readers may note that this is indeed not a formal proof. However, using special techniques and supercomputers we are yet to get a counter example. This led to the belief that the conjecture is correct. However, a formal proof will be a feather in the cap of Mathematician(s) — so let us wait and watch like the proof of Fermat's Last Conjecture.

```
#include <stdio.h>
#include <stdlib.h>
int main()
{
    int m,n, a, b,c,i,j, pc, primesN[200000], p1, p2, max=0, lmax=0, lc=0;
    int eratosthenesSIEVE(int , int []); /* computes and fills
```

```
                                  all the primes from
                                  2 to n, in the array pointed
                                  to by the 2nd parameter and
                                  and returns no. of primes */
  m = atoi(argv[1]);
  n = atoi(argv[2]);
  primesN[0] = 0;
  primesN[1] = 0;
  pc = eratosthenesSIEVE(n, primesN);
  c = m; /* From 4 to n specified in the command line */
  while (c <= n){
    for (i=2; i <= n/2; i++){
        if( primesN[i] == 0)
          continue;
      for(j = i; j < n; j++){
        if( primesN[j] == 0)
          continue;
        if (i+j > c) break;
        if ( primesN[i]  + primesN[j]   == c){
        lmax++; /* increase combination count */
          printf("%d =  %d +  %d\n ", c, primesN[i], primesN[j]);
        }
    }
  }
  }
if (max < lmax){
  max= lmax;
  lc =c;
  }
  c = c+2;
  lmax=0;
  }
  printf("In the range of %d to %d the number %d has
          a maximum of %d solutions\n = ", m, n, lc, max);
  return 1;
}
```

The program outputs all solutions in the specified range plus the even integer that has a maximum number of solutions. The following table shows the maximum number of solutions for a particular range.

Range	No. with maximumsolutions	No. of solutions
1. 4 to 10	10	2
2. 4 to 100	90	9
3. 4 to 1000	990	52
4. 4 to 10000	9240	329

7.4.1 Conjecture — got a formal proof

The most famous conjecture which has been ultimately proved is Fermat's Last conjecture. We know that Pierre de Fermat stated this proposition as a theorem about 1637 and had written that he had a proof that did not fit in the margin of his note-book. The theorem says $a^n + b^n = c^n$ is true for integer values of $n = 1$ and 2 only; obviously there is no solution for $n > 2$. After rigorous efforts by so many great minds, the proof came only in 1995. Prior to that, using computers it was checked that for prime numbers up to 4,000,000 the theorem holds good. Here we will show that for $n = 3$ to 10 and a, b varying from 1 to 15, Fermat's theorem complies with the prediction.

```c
int main()
{
   int i, j, k, n, status;
  for(n=3; n<=10; n++){
   for (i=1; i<= 15; i++)
    for(j=1; j < = 15; j++){
        s = power(i, n) + power(j,n);
        c = 1;
        status = 0;
        while (c <= 30){
           if ( s == power(c, n))
             printf("Counter example found\n");
             status = 1; break;
           c++;
      }
    }
     if ( status == 0)
         printf("No Counter example found\n");
   }
       return 0;
}
```

Testing with higher values of a, b and n would require techniques where numbers outside the range admissible with *long* or *long long* integers have to be treated as array of characters. We would see one such example for calculation of 100!. Incidentally, factorial of 21 onwards the results are very big numbers and beyond the range of integers in a typical PC.

7.4.2 Conjecture proved to be wrong

In line with Fermat's last theorem Euler conjectured that $\sum_{i=1}^{n-1} a_i{}^n \neq a_n^n$ for $n \geq 3$. Using computers several counter examples have been found including $27^5 + 84^5 + 110^5 + 133^5 = 144^5$. In the following program we have assumed that Goldbach's conjecture holds good for $n = 3$ and 4 and we are searching the possibilities of a counter example for $n = 5$.

```c
int main()
{
   int i, j, k, l, m;
```

```
long long s;
label done:
long long power(int , int);
for(i=1; i<=200; i++) /* The choice of i, j, k, ... etc. is dictated
                          by the already known result to reduce the
                          search space */
  for(j=1; j<=200; j++)
    for(k=1; k<=200; k++)
      for(l=1; l<=200; l++){
             m = 1;
             while ( m <= 300){ /* for n=5 we have tested it */
               s = power(i,5)+ power(j,5) + power(k, 5) + power(l, 5);
               if (s == power(m, 5)){
                    printf("Counter example with %d %d %d %d and %d\n");
                    goto done; /* goto may be used to get out from deep
                                    inside the loop in special cases */
               m++;
             }
      }
  }
done: return 0;
}
```

The statement(s) deep inside the loop will be executed 48×10^{10} times if there is no match. And even for a moderately powerful machine this would take days. So, special techniques have to be employed to search this vast solution space.

In order to get the output quickly you could reduce the trial space by using $i = 1$ to 50, $j = 80$ to 90; $k = 100$ to 120 and $l = 140$ to 150 as the result is already known. However, for unknown values even supercomputers with special programming techniques employed to deal with big numbers take substantial amount of time. In this context you may recollect that even with multiple supercomputers to find the next prime we have to wait years to get the next value of the prime.

We are now ready to solve some of the problems of Chapter 4. However, in the next section we provide a table classifying the type of the problems and hint to go either for a computer based solution or analytical solution.

7.4.3   Problems to be solved through computer programs

We now hope that an interested reader would be able to solve many problems using computer programs with the help of built-in functions and the functions that are presented here. We reiterate that analytical solution should be pursued in most of the cases. However, in parallel you may use a computer as well. For the benefit of programmers, we present a table here suggesting the problems (number mentioned in Chapter 4) that can be solved using simple programs. Through the same table we have provided a list of problems that are to be treated analytically. Moreover, hints are also given in the same table regarding the type of the problem and the corresponding functions to be used to solve it.

Table 7.1: Showing the type of the problems: Problems marked as analytical may not be suitable for solutions based on programs. Others can be solved with programs. Possible use of the functions is also indicated.

Type	Problem No.	Functions required	Comments
1. Simple Arithmetic	1, 21b, 21c, 21d, 22, 28, 28, 29a, 30, 31, 33, 37, 39, 41-47, 58, 59, 62, 63, 64, 66, 68b, 79, 87, 101, 112, 122, 123, 124, 128, 131, 132	Assorted or none	Can be solved using basic arithmetic operations and occasionally some built-in functions and functions developed in Chapter 7
2. Digits, Factors, Divisors, Primes, and Factorials	2, 7-12, 14, 16-19, 20, 21a, 24b, 24c, 25, 30, 32, 34, 39, 40, 42, 43, 48, 49, 50, 82, 85, 106, 108	Assorted functions developed in Chapter 7	
3. Random Numbers	51, 65, 89, 127, 134, 136	rand(), randomized()	
4. Big Numbers	13, 15, 23, 24, 26, 27, 29b, 33, 55, 56, 57, 64b, 69, 126, 129, 135	Big number functions like bigADD, bigSUB(), bigMUL() etc.	Numbers out of range are to be treated as character strings
5. Analytical	3-6, 38, 52, 53, 54, 59b, 67, 68a, 70, 71, 72, 73, 74, 75, 77, 78, 83, 84, 86, 88, 90, 91, 92, 93, 94, 95, 96, 97, 98, 99, 100, 102, 103, 104, 105, 107, 109, 110, 113, 114, 115, 116, 117, 118, 119, 120, 125, 130, 133, 137	None of the functions developed in Chapter 7 would help	Analytical solution is preferred. A few can be done through programs using a lot of analytical logic

7.5   Solutions to some of the problems

In this section we have provided solutions to some of the problems stated in Chapter 4. The problems are not chosen in increasing serial order for obvious reasons.

Problem No. 66

Here we compute x and y that satisfy $x^2 + 615 = 2^y$.

```c
#include <stdio.h>
int main()
{
    int x, y;
    int mypow(int , int);

    for(y=9; y<=100; y++)
      for(x=1; x<=100; x++)
        if (x*x + 615 == power(2, y))
            printf("x=%d, y=%d\n", x, y);
    return 0;
}
#include "power.c"
```

[Note that you may also use the *pow*() function — in that case include math.h and use the -lm (i.e. link math library switch) with the compile command]

Problem No. 24b and 24c

We compute the sum of all 3-digit palindromes and how many of them are not divisible by 11.

```c
#include <stdio.h>
int main()
{
    int i, sum=0; nodiv11=0;
    int getDIGIT(int , int);    /* Template for the function getDIGIT() */
    for(i=100; i<=999; i++){
        if ( getDIGIT(i, 1) == getDIGIT(i,3))
            sum = sum +i;
        if (i % 11 != 0)
            nodiv11++;
    }
    printf("Sum of all 3-digit palindroms is %d\n", sum);
     printf("No. of 3-digit palindroms not divisible by 11 is %d\n", nodiv11);
    return 0
}
#include "getDIGIT.c"
/* Assuming the function is separately
available as getDIGIT.c  in the folder */
```

Problem No. 48

Here we compute how many digits in total are used to write 1 to 100,000. In this program we may use the howmanyDIGITS() function effectively.

```c
#include <stdio.h>
int main()
```

```
{
    int i, sum=0;
    int howmanyDIGITS(int);
    for(i=1; i<=100000; i++)
      sum += howmanyDIGITS(i);
    printf("Total number of digits used to write 1 to 1000000 is %d\n", sum);
    return 0;
}
#include "howmanyDIGITS.c"
```

Problem No. 85

We find out the number of pages in a book starting with page number 1 and given that a total of 3189 digits have been used to print the page numbers

```
#include <stdio.h>
int main()
{
    int i,j, sum=0;
    int howmanyDIGITS(int);
    i=1;
    while(sum != 3189){
        sum += howmanyDIGITS(i);
        i++;
    }
    printf("No. of pages in the book is %d\n", i-1);
    return 0;
}
#include "howmanyDIGITS.c"
```

Problem No. 29a

Here we have to prove that no product of four consecutive counting numbers is a perfect square. The analysis is fairly easy and so is the programming counterpart. The program computes the product of four consecutive counting numbers like 1, 2, 3 and 4 to $n, n + 1, n + 2$ and $n + 3$ and in each case it checks if the product is a perfect square. Now the problem with the programming approach is that the product becomes very big and selecting n greater than 2000 is not possible due to the restriction of 64-bit arithmetic.

```
#include <stdio.h>
#include <math.h>
int main()
{
    long p, q, r, i, d, q2;
    r = 0; /* Perfect square count */
    for(i=1; i<=100; i++){ /* considering  100 sets of four
                                consecutive numbers */
        p = i*(i+1)*(i+2)*(i+3);
        q = (long) sqrt(p);
```

```
        if ( p - q * q == 0)
            printf("%ld is a perfect square\n", p);
    }
    return 0;
}
```

Since the problem statement is known to be true (proved arithmetically in chapter 5), we know that the *printf()* statement will not be executed at all. Now let us see what happens if you select n = 19XX. You will find that at least four products will be marked as perfect squares. The reason is with increasing *n* the product can no longer be contained in 64-bits and due to this overflow the product will be erroneous and may be equal to a perfect square. Later on we will see how to deal with very big integers by treating them as strings. However, for most of the programming problems overflow (or underflow) must be considered.

Problem No. 34

We compute the product of the digits of a natural number *n* and see if it satisfies the equation $n^2 - 15n - 27$.

```
#include <stdio.h>
int main()
{
    int i,t, p, pos;
    int getDIGIT(int , int);  /* Template for the function getDIGIT() */
    for(i=1; i<=999999; i++){ /* Testing natural number in this range */
        p = 1;
        t = i;
        pos = 1;
        while (t != 0){
            p = p * getDIGIT(i, pos);
            pos++;
            t = t/10;
        }
        if ( p == i*i -15*i - 27)
            printf("Here is a match for %d\n", i);
    }
    return 0;
}
#include "getDIGIT.c"
```

Problem No. 1

Here A and B vary from 1 to 9 and 0, 2, 4, 6, and 8, respectively. Note that the number is divisible by 72, an even number, so the number itself is an even number. So, the digit in the unit place takes 0, 2, 4, 6 and 8 only.

```
int main()
{
    int i, j, k;
```

```
      i = 1; j = 0;
      while (k <= 942738){
          k = i * 100000 + 42730 + j; /* k is a probable candidate
                                          A4273B -- A varies from
                                          1 to 9 and B varies from 0 to 8 */
          if ( k % 72 == 0)
              break; /* Exit from the loop */
          i = (i < 9) ? i++ : 1; /* equivalent to  if (i<9) i++; else i=1; */
          j = (j < 8) ? j+=2 : 0;
      }
      printf("%d\n", k);
      return 0;
}
```

Problem No. 2

Fortunately, we could calculate 20! keeping the result in a long variable. We also have digit extractor through which we could get the digits in 8th (i.e. x) and 7th (i.e. y) positions from the right.

```
main()
{
  printf("x = %d and y=%d\n", getDIGIT(fact(20), 8), getDIGIT(fact(20), 7);
  return 0;
}
```

Problem No. 3

An interesting division problem whose solution depends on your observation. Trying a blind search for a quotient whose digit in the hundred place is 8 (XX8XX) while the divisor varies from 100 to 999 (XXX) and dividend varies from 10000000 to 99999999 (XXXXXXXX) would result in 20762755 probable solutions. Which one is the correct solution? Now if you observe the way this problem was solved then you could easily guess that the last 4 digits of the quotient would be 0809. You could reduce the number of cases by checking each quotient fulfilling this criterion. Still you will have 20736 solutions. So, in such cases you are going indirectly for analytical solution as you have seen in chapter 5.

The program is given below that may be executed to see the number of cases.

```
#include <stdio.h>
#include "myfunctionheaders.h"
int main()
{
    int i, j, n, cn=0;
    int getdigit(int , int );
       for(i=10000000; i<=99999999; i++)
         for(j=100; j<= 999; j++)
           if (i%j == 0)
               if (getdigit(i/j,3 )== 8) cn++; /* Change this line to
```

```
                                    if ( (i/10000 == 0809) cn++;
                                    to reduce the no. of
                                    matches */
    printf("%d\n", cn);
    return 0;
}
```

Problem No. 25

Here we have a 6-digit number (abccba) with the condition that; b is odd and a, b and c may or may not be the same. So, the programming solution is:

```
int main()
{
    int i, j, c=0; int d[6]; /* to keep the digits of a number --
                                unit in d[0], 10th in d[1] ... */
    i = 110011;
    while (i <= 999999){
      for(j=6; j >=0; j--)
        d[j] = getDIGIT(i, j);
      if ( i%7 == 0 && d[0] == d[5] && d[2] == d[3]  && d[1] = d[5]
                    && d[1] % 2 == 1)
              c++;
      i++;
    }
    printf("%d\n", c);
    return 0;
}
```

With the help of the programming solutions provided here we hope that an interested reader may be able to solve more problems. In the next section we discuss how to deal with large integers.

7.6   Arithmetic with large integers

Readers may have already noticed that the range of numbers allowed to be represented in a typical computer is limited by the hardware. For example, if the ALU has 64-bit Full Adder then longest integer is also limited to 64-bit as well. For this reason integers are stored usually in 4-bytes and long (or long long) integers in 8-bytes. An unsigned integer stored in 8-bytes gives you a maximum of $2^{64} - 1$. Even this huge counting number fails to represent the factorial of 21!. One immediate solution is to represent even an integer as a real, getting a bigger range. However, the loss of precision that results by using real form may lead to unacceptable results. Fortunately, we may deal with large integers, henceforth referred to as *Big Numbers*, by treating them as an array of numeric digits each occupying 1-byte only. Regarding, the arithmetic operations, namely addition, subtraction, etc. the logic followed is the same except the operands are now digits in character (ASCII code) form and each one is to be treated separately. Let us explain it through a simple example adding two numbers A, and B, say 9507 and 567, respectively. Note that the numbers in the character array (string) forms would be the strings.

|9|5|0|7| and |8|6|7|, respectively.

Note that in memory actually the ASCII codes of the digits are really stored as shown below:

|57|53|48|55| and |56|54|55|, respectively.

The addition operation is shown below digit by digit from the rightmost following the usual steps for addition. For better comprehension we have shown the digits and not their true memory representation; i.e. ASCII codes. Moreover, to make the strings equal the smaller number is padded with 0's on the left.

Carry	1	0	1	0
Augend	9	5	0	7
Addend	0	8	6	7
Sum	0	3	7	4

Note the following
- The carry is 0 for the unit (rightmost) place. We have added 7 and 7 here in this unit place and got the sum 4 and carry for the next place (10th) as shown in the table.
- Two n-digit numbers when added may produce $(n + 1)$-digit result.
- The result to be considered is the carry and the sum together.
- In our case the leftmost column adding 9 and the carry 1 from the previous stage produces 0 as sum and carry as 1 for the next stage (not shown in the table) and the result is 10374.

7.6.1 Factorial of $n (\geq 21)$

It is easy to compute the factorial of a number $n < 21$, beyond which you cannot handle such a big integer even with *long* or *long long* data type, i.e. 8-byte representation. The following program shows how to compute factorial of n where n can be very large. Once again the clue is to treat the number as a character string.

```c
#include <stdio.h>
#include <string.h>
int factorial(int num);
int multiply(int a, int arr[], int lim);
int atoi(char *);
int main(int argc, char *argv[])
{
```

```
      int m, n;
      if (argc != 2){
      printf("Run this program with one integer input\n");
              return -1;
      }
      n = atoi(argv[1]);
      m= factorial(n);
      printf("\n No. of digits in %d! is %d\n", n, m);
      return 0;
}
int factorial(int num)
{
      int pc = 0;
      int arr[5500]; /* Increase array size for number > 1929 */
      arr[0] = 1;
      int cnt, a, lim = 1;
      for(a = 2; a <= num; a++)
      {
            lim = multiply(a, arr, lim);
      }
      printf("\n%d!= ", num);
      for(cnt = lim - 1; cnt >= 0; cnt--)
      {
            printf("%d", arr[cnt]);
    pc++;
      }
      return pc;
}
int multiply(int a, int arr[], int lim)
{
      int cnt, product, temp = 0;
      for(cnt = 0; cnt < lim; cnt++)
      {
            product = arr[cnt] * a + temp;
            arr[cnt] = product % 10;
            temp  = product / 10;
      }
      while(temp)
      {
            arr[lim] = temp % 10;
            temp = temp/10;
            lim++;
      }
      return lim;
}
```

The program may be executed using the following where a command line parameter (in our case 21) is passed to the program.

```
unix> ./a.out 21
```

The interested reader may see Appendix G to know about the parameters, namely, *argc* and *argv[]* and the concept of command line parameters.

Note that 21! = 51,090,942,171,709,440,00 is a 20-digit number. However, the largest unsigned integer that can be represented in 64-bits (8-bytes for a *long long* integer type in C) is $2^{64} - 1 = 18,446,744,073,551,615$ which is a 20 digit number but much smaller than 21!.

Appendix G contains the functions to add, subtract and multiply big numbers and with the clue presented above the reader, if interested, may easily comprehend the program logic. This addition of big numbers as character arrays, say n1[], and n2[] producing sum[] would take more time to compute and that is the trade-off to carry out big number manipulation in a typical programming environment. This gives you some clue on why not to expect announcement of the next new prime (an integer) in a couple of years though several supercomputers have been employed 24×7 to find them.

7.6.2 Solution to a Problems with big numbers

Problem no. 27 asks to sum up

$$6 + 66 + 666 + \cdots + 666\ldots66$$

where there are 6666 no. of 6's in the last term.

With the bigADD(n1, n2, sum) addition function we can solve this problem with the following program.

```
int main()
{
    char sum[7000], n1[7000]; /* maximum no. of characters is 6666,
                                 however in a blind approach we took
                                 more to accommodate carries */
    int i, j;
    for(i=0; i<7000; i++)
      sum[i]=n1[i] = 0;

    for(i=1; i<= 6666; i++){
      for(j=0; j<i; j++)
        n1[j] = '6';
      n[i] = '\0';
      addBIG(n1, sum, sum); /* sum[i] <-- sum[i] + n1[i]
                               simulating sum = 6 + 66 + ... + 666...66 */
    }

    :
    :
}
```

7.7 A computer cannot solve all problems

Computers are wonderful machine and offer immense logic and number crunching operations. Yet they cannot solve all the problems, even given any amount of time. One such problem is known as the *halting problem* in computing parlance — a discussion on which is beyond the scope of this book. Thus, in this section we would discuss some problems whose program implementation is easy, however, the results may not be available in years or even decades. Basically, these are the solution space explosion problems which cannot be handled in real-life terms.

7.7.1 Finding counter example for Euler's conjecture

Suppose, hypothetically, Euler's conjecture is true up to $n = 9$. You wrote a program, similar to the naive approach shown in this text, and expecting a counter example for $n = 10$. The equation to be tested for equality is given below:

$$a^{10} + b^{10} + c^{10} + d^{10} + e^{10} + f^{10} + g^{10} + h^{10} + i^{10} = j^{10}$$

It may not be hard to see that you need nine nested loops to vary, $a, b, c, \ldots, i$ from 1 to 1000 and you also need to have another loop to compare the sum to the 10th power of j which is, say, also varying from 1 to 1000. If we assume that we have a single instruction for comparison deep inside this 10th loop in our program then that will be executed $1000^{10} = 10^{30}$ times if the conjecture is true for $n = 10$. Now assume that a single instruction takes 0.001 nano-second (much faster than the average PC) or 10^{-12} second[2]. Then the computer would take $10^{30} \times 10^{-12}/(365 \times 24 \times 60 \times 60)$ or 31,709,791,984 years to carry out the comparison instruction alone.

7.7.2 The travelling salesman problem

Consider planning for a salesman covering all the places (say, cities), starting from the home city and coming back (a loop) without any repeat visit such that the cost (directly proportional to the distance between the cities) is minimum. The resource available is the location (the cities) and the distance between the cities. In mathematical terms the resource is a 2-dimensional array or matrix where the rows and columns are the cities and each element D_{ij} is the distance between city i and city j. A simple example with 4 cities is given below.

	A	B	C	D
A	0	4	21	13
B	4	0	45	19
C	21	45	0	12
D	13	19	12	0

[2]Intel Core i7 Extreme Edition 3960X score is 177,730 MIPS at 3.33 GHz, or 0.0056 ns/instruction

For *n* no. of cities you have $(n-1)!$ possibilities to be tested in a brute force approach. Using dynamic programming technique near optimum result can be obtained in a lesser time; however no polynomial time algorithm is available.

7.7.3 Cryptography — RSA (failure is the pillar of security)

In this digital era all transactions are made secure by using encrypted transmission of information by the sender which is decrypted at the receiving end. Compromising this security is not acceptable.

The idea of RSA is based on the fact that it is difficult to factorize a large integer. So, the failure to factorise is a blessing in disguise and RSA relies on it. RSA is an asymmetric cryptography algorithm that works on two different keys; public Key and private Key. The sender uses this public key for encrypted transmission of information; the cipher text. On the receiving side it is decrypted by the private key (see the figure below).

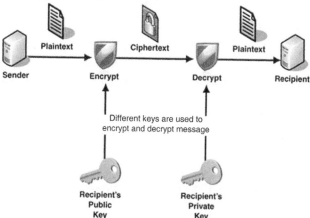

As the name suggests the public key is given to everyone and private key is kept private.

The transmission in encrypted form is done in three steps.
1. A client (for example a browser program on the client side) sends its public key to the server and requests some data.
2. The server encrypts the data using client's public key and sends the encrypted data.
3. Client receives this data and decrypts it using his private key.

In this digital era the transactional security is based on RSA or similar cryptographic techniques and it is imperative to emphasize its importance in our daily life.

The public key in RSA is derived by the multiplication of two large prime numbers. And the private key is also derived from the same two prime numbers. Though, technically factorization of this big number leads to a compromised private key and a failure of this scheme, but even the computer fails to do the factorization in any reasonable time and the security is practically impossible to breach. The encryption strength totally relies on the key size, and the strength of encryption increases exponentially if we represent the keys with more bits. RSA keys can be typically 1024 or 2048 bits long. Experts believe that 1024 bit keys could be broken in near future. But till now it seems to be an infeasible task.

Let us learn the mechanism behind RSA algorithm:

Generating public key

- Select two prime numbers P and Q. Suppose $P = 17$ and $Q = 23$.
- Now First part of the Public key, $n = P * Q = 391$.
- We also need a small exponent say e. This e must be an integer and not be a factor of n.
- Also, $1 < e < f(n)$ [$f(n)$ is computed as shown below].
- Let us now consider e be equal to 3. Our Public Key is made of n and e

Generating private key

We need to calculate $f(n)$: Such that $f(n) = (P - 1)(Q - 1)$ so, $f(n) = 352$. Now calculate Private Key, $d = (k * f(n) + 1)/e$ for some integer k. For $k = 2$, value of d is 235. Now we are ready with our public key ($n = 391$ and $e = 3$) and private key ($d = 235$).

Now, let us consider a message (data to be sent) 98:

- Thus Encrypted Data $c = 98^e$ mod $n = 55$
- Now we will decrypt 55:
- Decrypted Data $= c^d$ mod $n = 98$

Below is the C implementation of RSA algorithm.

```
/* C program for RSA algorithm.
 Small prime numbers and the message are used to avoid uisng strings
 representing big numbers */

#include<stdio.h>
#include<math.h>

int main() /* RSA */
{
    int gcd(int, int);
        /* Take two random prime numbers */
    double p = 3;
    double q = 7;

    double n = p*q;
    printf("1st prime=%g, 2nd prime=%g\n", p, q);

    double e = 2; /* Initial value for e */
    printf("Initial value of e=%g\n", e);
    double phi = (p-1)*(q-1);
```

```c
    /* Make e a co-prime to phi */
    while (e < phi)
      {
          if (gcd(e, phi)==1)
              break;
          else
              e++;
      }

    /* Now n and e together for the public key */

    printf("n=%g,  final value of e=%g\n", n, e);
    printf("n and final value of e form the public key\n");

    /* Private key d --
    choosing d such that it satisfies
    d*e = 1 + k * phi    */

    int k = 2;  /* A constant value is taken to multiple phi */
    double d = (1 + (k*phi))/e;
    printf("private key d=%g\n", d);

    /* Take a  message to be encrypted */
    double msg = 16;
    printf("Message data = %lf\n", msg);

    /* Encryption c = (msg^e) % n */

    double c = fmod(pow(msg, e), n);
    printf("\nEncrypted data = %lf", c);

    /* Decryption m = (c^d) % n */
    double  m = fmod(pow(c, d), n);

    printf("\nOriginal Message Sent = %lf\n", m);

    return 0;
}

/* Returns gcd of a and b */
int gcd(int a, int b)
{
    int t;
    while (1)
    {
        t = a%b;
        if (t == 0)
          return b;
        a = b;
        b = t;
    }
```

}

The output of the program is given below.

```
1st prime=3, 2nd prime=7
Initial value of e=2
n=21, final value of e=5
n and final value of e form the public key
private key d=5
Message data = 16.000000
Encrypted data = 4.000000
Original Message Sent = 16.000000
```

7.7.4   Difficult for analytical solution but easy for a computer

Time and again we harp on the importance of analysis and solving the problems by analysis. Here is a counter example.

The problem is to find out if the product of numbers a and b from a pool of 36 integers (1 to 36) is a perfect square or not and find out the total count of such perfect squares within this range.

Analytical reasoning leads us to the result of 27 such perfect squares. However, it is almost impossible to get the answer through analysis if we take three or four numbers at a time and try to find out if the product is a perfect square or not. The programming solution, on the other hand, is very simple.

```
#include  <stdio.h>
#inlude   <math.h>
int isPERFECTsquare(int n)
{
  if ( (int) sqrt(n) * sqrt(n) == n)
    return 1;
  else
    return 0;
}
int main()
{
    int a, b, p;
    for(a=1; a<=36; a++)
      for(b=a+1, b<=36; b++)
        for(c=b+1, c<=36; c++)
          for(d=c+1; d<=36; d++)
              if (isPERFECTsquare(a*b*c*d))
                  psc++;
          }
    printf("Total no. of perfect square is %d\n", psc);
    return 0;
    }
```

The program shown above is taking 4 numbers a, b, c and d at a time. It can be adopted for any range and any numbers to be taken at a time. The following table shows the result assuming appropriate changes are made in the program without changing the range from 1 to 36.

Numbers taken at a time → Total no. of perfect squares
2 → 27, 3 → 182, 4 → 1027, 5 → 4336, 6 → 15336 7 → 46359

7.8   Enhancing computing power through distribution

Referring to Euler's conjecture or similar problems — if we could distribute the computation as a unit to different people then we could get the results within a reasonable time. Let's take an example. India has approximately 1.4 billion people. A good percentage of this population uses mobile phone, say 1 billion people. Now if we could distribute a unique computation to each then the problem can be solved in real time. For Euler's conjecture blindly starting from 1 to 200 for each loop we have seen that the inner statements gets executed 48×10^{10} times for $n = 5$ — a humongous job. Now imagine each such computation is distributed to 1 Billion people. Then time requirement is scaled down 1 billion times. So, $n - 1$ persons who would say that they failed to match but the person luckily got the exact computation would send an affirmative answer. For problems with enormous search space many supercomputers are employed these days over a network to share the computing load and to get the result within a reasonable time.

7.9   Road ahead

The time has come to wrap up and look forward. We hope that we have raised your interest in numbers, their representation in different spheres of activities and the many properties that amaze us. Well, we have not introduced anything new other than making you confident about the numbers and playing with them. Your disinterested peers might wonder why you are wasting time with numbers and perhaps you are trying to reinvent the wheel obviously without any tangible benefits. The simple answer to this question could be you are doing it for fun and for that lovely feeling "wow! I also can do it". In this context it may not be out of place to refer to the famous question asked to a renowned mountaineer, George Mallory — what is the reason for trying to scale Mount Everest? It is so dangerous and life threatening a pursuit, apparently without anything in return. We know the prompt answer was "because it is there". Numbers are all around us, so are some of the fascinating problems and puzzles. Why not give a try to scale them as "it's there."

Appendix A

Brahmagupta's Equation

A.1 Solution of Equation (1.7)

Let us first rewrite eq. (1.7) as shown below:

$$m^2 = \frac{n(n + 1)}{2} \tag{A.1}$$

with m and n as integers.

Now multiplying both sides by 8 and adding 1, we get

$$2(2m)^2 + 1 = (2n + 1)^2 \tag{A.2}$$

Substituting

$$p = 2n + 1, \quad q = 2m \tag{A.3}$$

in (A.2), one obtains

$$p^2 - 2q^2 = 1 \tag{A.4}$$

Equations of the form

$$p^2 - Dq^2 = 1 \tag{A.5}$$

where D is a non-square integer and p and q are integers, are now known as *Brahmagupta's equation*. In the west, this type of equation is more popularly known as *Pell's equation*. It may be noted that with D a perfect square, say, m^2, (A.5) reduces to $p^2 - (mq)^2 = 1$, implying two perfect square integers differing by 1. That equation obviously has only a trivial solution $p = 1$, $q = 0$.

If the values of the smallest set (p, q), say, (p_1, q_1) are somehow determined, then all the infinite set of values can be easily determined. Before showing how this can be done, it must be mentioned that the set of smallest values may fluctuate wildly with a little change in the value of D. This is illustrated below considering three consecutive values of D.

196

$$D = 60 \quad p_1 = 31 \qquad\qquad\qquad q_1 = 4$$
$$D = 61 \quad p_1 = 1{,}766{,}319{,}049 \qquad q_1 = 226{,}153{,}980$$
$$D = 62 \quad p_1 = 63 \qquad\qquad\qquad q_1 = 8$$

Brahmagupta (598–670) was aware of this phenomenon. So he declared anyone solving the equation $p^2 - 92q^2 = 1$, within one year must be a good mathematician. In this case, the values are $p_1 = 1151$, $q_1 = 120$. Brahmagupta found an algorithm to obtain this minimum set for (A.5), which was later improved by other Indians in the tenth and twelfth centuries. The Europeans independently found such methods in seventeenth and eighteenth centuries.

For some low values of D, like 2, 3 and 5, we can by inspection find the minimum sets as written below:

$$D = 2 \quad p_1 = 3 \quad q_1 = 2$$
$$D = 3 \quad p_1 = 2 \quad q_1 = 1$$
$$D = 5 \quad p_1 = 9 \quad q_1 = 4$$

Now let us return to (A.4) and discuss how to get all the values of (p, q) from the set $(p_1 = 3, q_1 = 2)$. Towards this goal, first we write (A.4) using the smallest set as

$$(3 - 2\sqrt{2})(3 + 2\sqrt{2}) = 1 \tag{A.6}$$

Squaring both sides of (A.6), we get

$$(17 - 12\sqrt{2})(17 + 12\sqrt{2}) = 1 \tag{A.7}$$

Writing (A.4) as $(p - \sqrt{2}q)(p + \sqrt{2}q) = 1$, it is easy to conclude that $(p_2 = 17, q_2 = 12)$. Proceeding in this manner, we finally obtain

$$p_k + \sqrt{2}q_k = (3 + 2\sqrt{2})^k, \quad k = 1, 2, 3, \ldots \tag{A.8}$$

All the values of (m, n) can be obtained using (A.3) and the values of (p_k, q_k). The reader should verify that these values come out as $(1,1),\ldots, (6, 8), (35, 49), (204, 288), \ldots$ as given in Chapter 1.

At this stage we may just point out a connection between the infinite continued fraction of Ramanujan, mentioned in Chapter 1, with the solution given above. Let the infinite continued fraction of Ramanujan be given by x, i.e.

$$x = \cfrac{1}{6 - \cfrac{1}{6 - \cfrac{1}{6 - \ldots}}} = \frac{1}{6 - x}$$

Or, $x^2 - 6x + 1 = 0$, taking only the root < 1, we get $x = 3 - 2\sqrt{2}$, when $\frac{1}{x} = 3 + 2\sqrt{2}$. Thus, $(3 - 2\sqrt{2})(3 + 2\sqrt{2}) = 1$, which is same as the generating eq. (A.6) from where all the values were obtained!

A.2 Solution of Equation (2.29)

First we reproduce eq. (2.29) as

$$a^2 - 2z^2 = -1 \tag{A.9}$$

We follow a procedure similar to what is explained in Appendix A.1. From inspection it is easy to find the smallest solution as $(a_1 = 1, z_1 = 1)$, when (2.28) can be written as

$$(1 - \sqrt{2})(1 + \sqrt{2}) = -1 \tag{A.10}$$

As explained in Section A.1, it is easy to see that by raising both sides of (A.10) to odd powers, we get

$$a_n + \sqrt{2}z_n = (1 + \sqrt{2})^n, \quad n = 1, 3, 5, \ldots \tag{A.11}$$

Using various values of n, we get the desired Pythagorean triples as given below:

$n = 3$, $a_3 = 7$, $z_3 = 5$; when $x_3 = (7 - 1)/2 = 3$ and $y_3 = (7 + 1)/2 = 4$

$n = 5$, $a_5 = 41$, $z_5 = 29$; when $x_5 = (41 - 1)/2 = 20$ and $y_5 = (41 + 1)/2 = 21$

and so on.

Appendix B

Bernoulli Numbers

Sum of integral powers of consecutive natural numbers

Let us first reproduce eq. (3.19)

$$S^{(k-1)} = 1^{k-1} + 2^{k-1} + 3^{k-1} + \cdots + n^{k-1}$$

$$= \frac{1}{k} \left[B_0 n^k + B_1 \binom{k}{1} n^{k-1} + B_2 \binom{k}{2} n^{k-2} + B_3 \binom{k}{3} n^{k-3} + \ldots \right] \qquad (B.1)$$

With $k = 1$ in (B.1), we get

$$S^{(0)} = 1 + 1 + 1 + \cdots + 1 = n = \tfrac{1}{1}(B_0 n) \qquad (B.2)$$

with $B_0 = 1$.

Now with $k = 2$ in (B.1), we write

$$S^{(i)} = 1 + 2 + 3 + \cdots + (n - 2) + (n - 1) + n$$

which we calculate as follows.

First we write

$$n^2 - (n - 1)^2 = 2n - 1$$
$$(n - 1)^2 - (n - 2)^2 = 2(n - 1) - 1$$
$$(n - 2)^2 - (n - 3)^2 = 2(n - 2) - 1$$
$$\vdots$$
$$2^2 - 1^2 = 2.2 - 1$$
$$1^2 - 0^2 = 2.1 - 1$$

Adding all the above equations, we make a telescopic sum to get

$$n^2 = 2S^{(1)} - S^{(0)}$$

$$S^{(1)} = \tfrac{1}{2}(n^2 + n) = \tfrac{1}{2}\left(B_0 n^2 + B_1 \binom{2}{1} n\right) \qquad \text{(B.3)}$$

with $B_0 = 1$, $B_1 = \tfrac{1}{2}$.

Now we calculate $S^{(2)} = 1^2 + 2^2 + 3^2 + \cdots + (n-2)^2 + (n-1)^2 + n^2$ in the way explained below. First we write

$$n^3 - (n-1)^3 = 3n^2 - 3n + 1$$
$$(n-1)^3 - (n-2)^3 = 3(n-1)^2 - 3(n-1) + 1$$
$$(n-2)^3 - (n-3)^3 = 3(n-2)^2 - 3(n-2) + 1$$

$$\vdots$$

$$2^3 - 1^3 = 3.2^2 - 3.2 + 1$$
$$1^3 - 0^3 = 3.1^2 - 3.1 + 1$$

The telescopic sums of all the above equations yields

$$n^3 = 3S^{(2)} - 3S^{(1)} + S^{(0)}$$

$$S^{(2)} = \tfrac{1}{3}(n^3 + 3S^{(1)} - S^{(0)}) = \tfrac{1}{3}\left(n^3 + \tfrac{3}{2}n^2 + \tfrac{3}{2}n - n\right)$$
$$= \tfrac{1}{3}\left(B_0 n^3 + B_1 \binom{3}{1} n^2 + B_2 \binom{3}{2} n\right) \qquad \text{(B.4)}$$

with $B_0 = 1$, $B_1 = \tfrac{1}{2}$, $B_2 = \tfrac{1}{6}$.

A similar procedure can be followed by increasing the value of k by 1 at every step to get the formula for any value of k. The reader is advised to note that the entries of the kth row in the Pascal's triangle (except the last 1) in the expressions of $S^{(k)}$ for every value of k to define accordingly the Bernoulli numbers B_k's.

Appendix C

Unit Fractions

Prime numbers and unit fractions

In Section 3.8.2, we mentioned that the reciprocal of a prime number p can be expressed in a unique way as the difference of two unit fractions as given below:

$$\frac{1}{p} = \frac{1}{p-1} - \frac{1}{p(p-1)} \tag{C.1}$$

whereas for a composite number C its reciprocal can be expressed in multiple ways as the difference of two unit fractions. We discuss below the general proof of this statement and also show for a composite number how many ways it can be done and how to obtain those unit fractions.

First let us consider (C.1) corresponding to any natural number n instead of a prime number p and write its reciprocal as the difference of two unit fractions as

$$\frac{1}{n} = \frac{1}{i} - \frac{1}{j} \tag{C.2}$$

It is clear that in (C.2), $n > i$ and $j > i$. Let $i = n - m$ with m representing natural numbers. Thus,

$$\frac{1}{n} = \frac{1}{n-m} - \frac{1}{j} = \frac{j-n+m}{j(n-m)} \text{ or, } j(n-m) = n(j-n+m)$$

$$\text{or, } n^2 = m(n+j) \tag{C.3}$$

It is known that $m \le n - 1$ and if n is a prime p, then n^2 has only one factor $\le p - 1$, that is 1; since it has only three factors $1, p$ and p^2.

So the only possible value of m is 1. Thus, finally we get (C.1) from (C.2) with $i = n - 1 = p - 1$, which immediately gives $j = p(p-1)$.

Now consider that n is a composite number, where $n = rs$, with $n > 1, r > 1$.

201

Now (C.3) can be written as

$$r^2 s^2 = m(rs + j) \tag{C.4}$$

where $1 \leq m \leq rs - 1$.

Thus, we see m, a divisor of n^2 must be $\leq (n - 1)$ and all such values of m yield a solution as claimed in Section 3.8.2.

Appendix D

Pattern in Decimal Representation

Derivation of Curious pattern shown in section 3.8.2

We start with the well known sum of an infinite GP series of the form

$$1 + x + x^2 + x^3 + \cdots = \frac{1}{1-x} \tag{D.1}$$

where $|x| < 1$.

Squaring both sides of the above equation, we get

$$1 + 2x + 3x^2 + 4x^3 + \cdots = \frac{1}{(1-x)^2} \tag{D.2}$$

The lefthand side of (D.2) can be easily obtained by carrying out the squaring process as explained below. Write (D.1) in tabular form, Table D.1, while writing the terms appearing on the lefthand side both in the first row and first column. Then write the product of the first row by each element of the column to get the subsequent rows. Then starting from the second row and second column, collect all the terms diagonally (i.e. terms having the same power of x) to get the righthand side of (D.2).

Table D.1

$1/(1-x) =$	1	x	x^2	x^3	x^4	$\cdots$
1	1	x	x^2	x^3	x^4	$\cdots$
x	x	x^2	x^3	x^4	x^5	$\cdots$
x^2	x^2	x^3	x^4	x^3	x^6	$\cdots$
x^3	x^3	x^4	x^5	x^6	x^7	$\cdots$
x^4	x^4	x^5	x^6	x^7	x^8	$\cdots$
$\vdots$	$\cdots$	$\cdots$	$\cdots$	$\cdots$	$\cdots$	$\cdots$

Substituting $x = 0.1$ in (D.2), we obtain $1 + 0.2 + 0.03 + 0.004 + \cdots = 100/81$.

Dividing both sides by 100, it is easy to see that $0.01234567901234\cdots = 1/81$.

Attention may be drawn that due to carryover in the summing process, the digit 8 has been converted to 9 and consequently the digit 8 is missing. The reader is advised to carry out the same procedure by putting $x = 0.01$ and 0.001 in (D.2) to obtain the curious repeating patterns of the digits in the decimal expression of 1/9801 and 1/998001 reported in Section 3.8.2.

Appendix E

Platonic Solids

Platonic solids, golden rectangle and Euler's formula

Greek geometers proved that there can only be five convex polyhedrons which are bounded by identical regular polygons. All these five are shown in Figs E.1(a)–(e); these are known as *ideal Platonic solids*. Table E.1 shows the names, which are based on the number of faces. The faces can only be equilateral triangles, squares or regular pentagons. In the Table E.1 the number of vertices (V), edges (E) and faces (F) are also included for each polyhedron. It can be seen each of them satisfies Euler's celebrated formula

$$V - E + F = 2 \qquad (E.1)$$

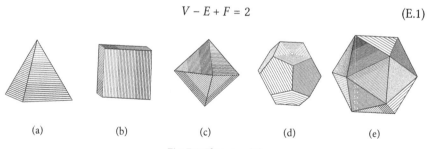

(a)   (b)   (c)   (d)   (e)

Fig. E.1: Platonic solids

The reader is advised to imagine and draw a hexahedron with equilateral triangles as its faces. Also verify that for this solid $V = 5$, $E = 9$ and $F = 6$ satisfying (E.1).

Table E.1

Fig. D.1	Name	V	E	F
(a)	Tetrahedron	4	6	4
(b)	Hexahedron	8	12	6
(c)	Octahedron	6	12	8
(d)	Dodecahedron	20	30	12
(e)	Icosahedron	12	30	20

As mentioned in Section 3.9.1, the Greek geometers were obsessed with golden quantities. They searched for such golden things in almost anything they thought are of ideal shape. As a result, they could locate golden rectangles in two of these Platonic solids. Figs E.2(a) and (b) show the presence of three golden rectangles in each of these two solids. We may notice that in a dodecahedron, there are 12 faces and we choose three groups each having a set of four faces. Consider the centroids of each of these regular pentagonal faces. Joining the centroids three golden rectangles have been obtained. Similarly in the icosahedrons consisting of 12 vertices, we choose three groups each having a set of four vertices. Joining these vertices, one again gets the golden rectangles.

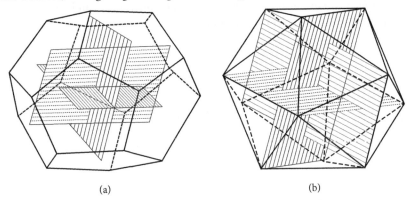

(a)   (b)

Fig. E.2: Golden rectangles in dodecahedron and icosahedron

Appendix F

Conversion table, Boolean Algebra rules, Floating Point Format etc.

F.1   Decimal to Binary, Octal and Hexadecimal

Decimal	Binary	Octal	Hex	Decimal	Binary	Octal	Hex
0	0	0	0	1	1	1	1
2	10	2	2	3	11	3	3
4	100	3	4	5	101	5	5
6	110	6	6	7	111	7	7
8	1000	10	8	9	1001	11	9
10	1010	12	A	11	1011	13	B
12	1100	14	C	13	1101	15	D
14	1110	16	E	15	1111	17	F

F.2   Laws of Boolean Algebra and De-Morgan's Laws

Commutative	Associative	Distributive
$A + B = B + A$ $A . B = B . A$	$A + (B + C) = (A + B) + C$ $A * (B * C) = (A * B) * C$	$A + (B + C) = (A + B) + C$ $A . (B + C) = A . B + A . C$
Annulment	Identity	Idempotent
$A + 1 = 1$ $A . 0 = 0$	$A + 0 = A$ $A . 1 = A$	$A + A = A$ $A . A = A$
Complement	Double Negation	Absorption
$A + A' = 1$ $A . A' = 0$	$(A')' = A$	$A + AB = A$ $A . (A+B) = A$
De Morgan's 1st Law	De Morgan's 2nd Law	Comments on De Morgan's
$(A . B)' = A + B'$	$(A + B)' = A' . B'$	Useful for logic synthesis

F.3 FLOATING POINT: Normalised, De-normalised and NaN

With reference to the IEEE 754 format given in the main text the following have been added for more depth in your understanding of floating point representation.

Depending on the bit string in the exponent part (e) the real numbers stored using IEEE 754 are interpreted as (i) Normalised numbers; the common case, (ii) Denormalised number and (iii) Special values; like (a) Infinity (b) Not a Number (NaN).

Normalised	s	e = 1 ... 254	f
De-normalised	s	e = 0000 0000	f
Infinity	s	e = 1111 1111	f = 000 0000 0000 0000 0000 0000
NaN	s	e = 1111 1111	$f \neq 0$

- **Normalised numbers**
    - e can be any 8-bit value except 0 and 255.
    - Consequently for a bias b of 127 $E = e - b$ will have the range of -126 to 127.
    - f represents any 23-bit string and the significand is normalised by adding 1.0 with f giving $1 \leq M < 2$ range to M.
- **Denormalised number**

Denormalized number is a requirement as through the normalised representation you.
    - Cannot represent zero; and
    - small numbers close to zero may have less accuracy.

The e bits would be zero indicating a **De-normalised** representation for which
    - $E = 1 - b$; and
    - $M = 0.f$; i.e. 1.0 would not be added and M will have the range of $0 \leq M < 1$.

f represents any 23 bit string for single precision.
- **Special and NaN**

This is a clever way of representing non-real numbers or something which are exceptional cases.

Here, e bit string consists of all 1s. And we have two different cases.
    - If the f part is zero then the representation indicates $+\infty$ or $-\infty$ depending on the sign bit s.
    - If the f part is non-zero then the representation indicates any NaN indicating something which is not a real number. Say, for example $\sqrt{-1}$ or

a real variable not yet assigned any legitimate value or 'divide by zero' error by using some agreed upon pattern of f bit string.

F.3.1 Examples of Floating Point Encoding: Normalized

As a practical example let us examine how an integer x = 12345 is encoded if we change it to FP (i.y., float y = (float) x;)

The binary encoding of 12345_{10} = 11 0000 0011 1001. We create a normaized representation of the above by

$$12345_{10} = 1.\textbf{100 0000 1110 01} \times 2^{13}$$

[*Note*: The binary point is moved towards left by 13 position; so to keep the value same in normalized form we need to multiply it by 2^{13}.]

So, the final form in single precision is

$$0\ 10001100\ \textbf{100 0000 1110 01}\ 0000000000$$

[*Note*: MSB is s = 0; e = 140. Therefore, $E = e - 127$ (the bias) = 13; reflecting multiplication of the normalized M by 2^{13}. Also see that the ten 0's in the LSB position to make f to 23-bits. Finally, the leading 1 of M is not stored.]

F.3.2 Examples of Encoding: De-Normalized and NaN

Number	32-bit Hex (single precision)	Type
0.0	00000000	De-normalized
−0.0	80000000	De-normalized
$\sqrt{-1}$	7FC00000	NaN
$-\sqrt{-1}$	FFC00000	NaN
$+\infty$	7F800000	Special
$-\infty$	FF800000	Special

Float variable x is assigned different values and the corresponding memory storage is displayed in Hex to see the encoding for comprehension.

Let us see the encoding of a normalized negative real number, say x = −12345.75. The binary encoding is

-12345.75_{10} = −11 0000 0011 1001.11. Note, the presence of 11 after the binary point (.). The corresponding Floating point is given below:

$$1\ 10001100\ \textbf{100 0000 1110 01}\ 11\ 00000000$$

[*Note*: MSB is s = 1; e = 140. Therefore, $E = e - 127$ (the bias) = 13; reflecting multiplication of the normalized M by 2^{13}. Also see (i) the presence of 11 (before the trailing eight zeroes that reflects) $2^{-1}(= 0.5) + 2^{-2}(= 0.25) = 0.75$ and; (ii) the eight 0's in the LSB position to make f to 23-bits. Finally, the leading 1 of M is not stored]

F.3.3 Range

It may be noted that the difference between a 32-bit integer representation and the corresponding floating point representation is

- the larger dynamic range for the latter;
- all the integers are equally spaced in the number line;
- the real numbers are not equally spaced in the number line;
- Most of the real numbers cannot be exactly represented (we have infinite number of real numbers between any two real numbers however close); only approximated.

The following table shows some of the positive number representations in single precision.

Number	exp	frac	Value	Decimal
Zero	$00\ldots00$	$0\ldots00$	0	0.0
Smallest (Denorm.)	$00\ldots00$	$0\ldots01$	$2^{-23} \times 2^{-126}$	1.4×10^{-45}
Largest (Denorm.)	$00\ldots00$	$1\ldots11$	$(1 - \epsilon) \times 2^{-126}$	1.2×10^{38}
Smallest (Norm.)	$00\ldots01$	$0\ldots00$	1×2^{-126}	1.2×10^{-38}
One	$01\ldots11$	$0\ldots00$	1×2^0	1.0
Largest (Norm.)	$11\ldots10$	$1\ldots11$	$(2 - \epsilon) \times 2^{127}$	3.4×10^{38}

F.3.4 Rounding

Rounding is done to reduce inaccuracy in representing a real number. Thus, the goal to represent a number x is to find out the closest match x' in a systematic manner.

Instead of a close match, methods to set lower (x^-) and upper bounds (x^+) such that $x^- \leq x \leq x^+$ are also used.

There are 4 rounding methods used in IEEE 754 standard. The first one is to find a close match. The next three are based on setting a lower and an upper boundary. Here is a table for rounding monetary amount.

Mode	Values to be rounded				
	2.40	2.60	2.50	3.50	−2.50
Round-to-even	2	3	2	4	−2
Round-toward-zero	2	2	2	3	−2
Round-down	2	2	2	3	−3
Round-up	3	3	3	4	−2

F.4   ASCII Table

Decimal	Hex	Char	Decimal	Hex	Char	Decimal	Hex	Char	Decimal	Hex	Char
0	0	[NULLL]	32	20	[SPACE]	64	40	@	96	60	`
1	1	[START OF HEADWG]	33	21	!	65	41	A	97	61	a
2	2	[START OF TEXT]	34	22	"	66	42	B	98	62	b
3	3	[END OF TEXT]	35	23	#	67	43	C	99	63	c
4	4	[END OF TRANSMISSION]	36	24	$	68	44	D	100	64	d
5	5	[ENOUIRY]	37	25	%	69	45	E	101	65	e
6	6	[ACKNOWLEDGE]	38	26	&	70	46	F	102	66	f
7	7	[BELL]	39	27	'	71	47	G	103	67	g
8	8	[BACKSPACE]	40	28	(	72	48	H	104	68	h
9	9	[HORIZONTALTAB]	41	29	)	73	49	I	105	69	i
10	A	[LINE FEED]	42	2A	*	74	4A	J	106	6A	j
11	B	[VERTICAL TAB]	43	2B	+	75	4B	K	107	6B	k
12	C	[FORM FEED]	44	2C	,	76	4C	L	108	6C	l
13	D	[CARRIAGE RETURN]	45	2D	-	77	4D	M	109	6D	m
14	E	[SHIFT OUT]	46	2E	.	78	4E	N	110	6E	n
15	F	[SHIFT IN]	47	2F	/	79	4F	O	111	6F	o
16	10	[DATA LNKE SCAPE]	48	30	0	80	50	P	112	70	p
17	11	[DEVICE CONTROL 1]	49	31	1	81	51	Q	113	71	q
18	12	[DEVICE CONTROL 2]	50	32	2	82	52	R	114	72	r
19	13	IDEVICE CONTROL 3]	51	33	3	83	53	S	115	73	s
20	14	IDEVICE CONTROL 4]	52	34	4	84	54	T	116	74	t
21	15	[NEGATIVE ACKNOWLEDGE]	53	35	5	85	55	U	117	75	u
22	16	[SMCHRONOUS IDLE]	54	36	6	86	56	V	118	76	v
23	17	[ENG OF TRANS.BLOCK]	55	37	7	87	57	W	119	77	w
24	18	ICANCEL]	56	38	8	88	58	X	120	78	x
25	19	[END OF MEDIUM]	57	39	9	89	59	Y	121	79	y
26	1A	[SUBSTITUTE]	58	3A	:	90	5A	Z	122	7A	z
27	1B	[ESCAPE]	59	3B	;	91	5B	[	123	7B	{
28	1C	[FILE SEPARATOR]	60	3C	<	92	5C	\	124	7C	|
29	1D	IGROUP SEPARATOR]	61	3D	≡	93	5D	]	125	7D	}
30	1E	[RECORD SEPARATOR]	62	3E	>	94	5E	^	126	7E	~
31	IF	[UNIT SEPARATOR]	63	3F	?	95	5F	_	127	7F	[DEL]

F.5 ASCII version of hello.c

#	i	n	c	l	u	d	e	SP	<	s	t	d	i	o	.
35	105	110	99	108	117	100	101	32	60	115	116	100	105	111	46

H	>	\n	\n	i	n	t	SP	m	a	i	n	(	)	\n	{
72	62	10	10	105	110	116	32	109	97	105	110	40	41	10	123

\	SP	SP	SP	SP	p	r	i	n	t	f	{	"	h	e	l
10	32	32	32	32	112	114	105	110	116	102	40	34	104	101	108

l	o	,	SP	w	o	r	l	d	\	n	"	)	;	\n	SP
108	111	44	32	119	111	114	108	100	92	110	34	41	59	10	32

SP	SP	SP	r	e	t	u	r	n	SP	0	;	\n	}	\n
32	32	32	114	101	116	117	114	110	32	48	59	10	125	10

Below each character in a row the corresponding ASCII value in decimal is shown. SP indicates a space character.

F.6 C keywords

auto	break	case	char	continue
default	do	double	else	entry
extern	float	for	goto	if
int	long	register	return	short
sizeof	static	struct	switch	typedef
union	unsigned	void	while	

Appendix G

Your own Library, command line parameters and dealing with big numbers

G.1 Your own library

With a growing number of useful functions to be used and shared, creating a library is a prudent choice to keep the functions under a single head; that is a library. Basically, a library is nothing other than a collection of object modules (functions after compilation known as linkable object files/modules). A directory is added on top of this collection of objects files for faster searching. These object files of frequently used functions in a user defined library (your own library) can be linked with any application. A library can be statically or dynamically linked with other object files. Here we will discuss creation of static libraries using the archive (*ar*) command in a Unix system.

Assume that all functions are written in separate source files. For example, the source files for the functions oddOReven() and getFACTORS() are oddOReven.c and getFACTORS.c, respectively. The corresponding object files, oddOReven.o and getFACTORS.o, may be created using the following commands.

```
unix>gcc  -c oddOReven.c  /* Option -c  is used to create oddORevn.o */
unix>gcc  -c getFACTORS.c  /*  getFACTORS.o */
```

Mutiple object files can also be created using a single -c switch as shown below.

```
unix>gcc  -c  oddOReven.c getFACTORS.c
```

In the next step you have to create your library and putting the object files (.o files) using the following command.

```
unix>ar  rcs myfunLIBRARY.a oddOReven.o getFACTORS.o
```

The command is straightforward, use the appropriate options, *rcs* in our case, followed by the name of the library that you have chosen (in this case myfun-LIBRARY) with **.a** as an extension indicating it is an archive. This follows the name(s) of the already created object modules that you would like to insert in the library.

Later on, new functions can be added to this library as and when required. Suppose, a new function new.c is written and the linkable new.o is created as before. You can add this function to the already existing library by the same command as shown below.

```
unix>ar  rcs myfunLIBRARY.a new.o
```

The library may be linked with your main program so that functions used in the main program can be attached to the final executable object file. Here is a full example for the task of finding the odd and even factors of the number 240.

```
#include <stdio.h>

int main()
{
    int n=240, fac[20];
    int i, nf; /* No. of factors */

    int oddOReven(int );  /* function template */
    int getFACTORS(int, int []); /* function template */

    nf = getFACTORS(n);
    for(i=0; i<nf; i++)
        if (oddOReven(fac[i]) == 0)
            printf("The factor %d is EVEN\n", fac[i]);
        else
            printf("The factor %d is ODD\n", fac[i]);
      return 0;
}
```

The program is to be converted to a linkable object file. Assuming that the source file name is **mymain.c** the following command creates the object file **mymain.o**.

```
unix>gcc -c   mymain.c
```

Finally, link the object files to create a single executable fully linked file by using the following command.

```
unix>gcc -c   -static  -o mymain  mymain.o ./myfunLIBRARY.a
```

The option *-static* indicates you are linking (functions or object modules from) a static library named *myfunLIBRARY.a* with the object file *mymain.o* and creating an executable file named *mymain* (any other name of your choice like *myprog* can also be chosen).

You may have already noticed that if a function is defined after the main() module or elsewhere, the function templates need be added to your main() module (or in the body of a function which is calling another). Here comes the use of the header files. Instead of writing function templates in all of your main() and subsequent functions create a header file and include it just like we did it for *stdio.h* in all our programs. So, our header file *myfunHEADER.h* might have the following template lines.

```
int oddOReven(int );
int isPRIMES(int );
int getFACTORS(int , int []);
 .
 .
 .
```

The linking of libraries is shown in the following figure for better comprehension.

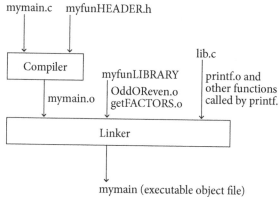

Now you need not write the templates separately in all of your programs, rather include them in the very beginning. For example

```
#include <stdio.h>
#include "myfunLIBRARY.h"

int main()
{
 /* No templates for the functions to be referred here */
 /* main program body here */
}
```

Note that *stdio.h* is a system supplied library and coming from a particular place known to the system. In your case the library may be in your directory where you write all your programs. This is reflected by the use of *myfunHEADER.h* unlike *<stdio.h>*.

G.2 Command line parameters

The main function in a C program accepts different parameters; *argc* and *argv* are frequently used to pass imporatant parameters to the main function when the program is invoked as a command. The parameter *argc* is an integer that counts the number of words in the command-line while *argv* is an array of pointers each pointing to the beginning of each word in the command-line. This helps avoiding execution time dialogue between the user and the program and a smart way to exit from a program if it is invoked with extra (or fewer) inputs. Suppose, we run a program findfact.c that computes factorial of a number *n*.

```
\include <stdio.h>
\include <stdlib.h>

int main(int argc, char *argv[] ){
   int fact(int);
   int ascii2integer(char [])
   int n;
   if ( argc != 2){
      printf("This program needs an integer as input in the command line\n");
      exit(0);
   }
   n = ascii2integer(argv[1]); /* argv[i] points to command line strings
                                  (words) so we need to convert the string
                                  to corresponding integer */
   printf("%d\n", fact(n));
   return 0;
}

int fact(int n)
{
    if (n==0 || n == 1)
       return 1;
    else
       return n * fact(n-1);
}

int ascii2integer(char s[])
{
   int n=0;
   for (int i=0; s[i] != '\0'; i++)
      n = n * 10 + s[i] - '0';
   return n;
}
```

This *findfact.c* program is compiled and the executable file, namely, *findfact* is run to find out the factorial of 12, used as a command line parameter, as shown below.

```
unix>findfact   12
```

Note that we have got two words 'findfact' and '12' in the command line; thus *argc* is 2. On the other hand argv[0] points to the beginning of the word 'findfact' and argv[1] points to the beginning of the word '12'. Note that 12 is stored in memory as 31H, 32H and \null in three consecutive bytes in the memory.

G.3   Functions for big numbers

A number of functions to deal with the big number are included here.

```c
#include <stdio.h>
#include <string.h>
/*    ADDITION of BIG numbers */
int bigADD(char n1[], char n2[], char sum[]) /* returns carry */
{
    int l1, l2; /* length counter */
    int i, j, k, c, t, d;

    void insert(char[], int);
    l1 = strlen(n1); l2 = strlen(n2); /* get the length of the strings */
    d = l1 - l2; /* see the difference  and make them same */
    printf("%d %d %d\n", l1, l2, d);
    if ( d >0) { /* n1[] is bigger-- add d no. of 0's in front of n2[] */
        insert(n2, d); l2 = l2+d;
    }
    else if ( d < 0) { /* n2[] is bigger -- add d no. of 0's in front of n1[] */
        insert(n1, -d);
        l1 = l1-d;
    }

/* strings are adjusted; l1 = l2; now add digitwise from the right */
    printf(" %s\n", n1);
    printf(" %s\n", n2);
    i = l1-1;
    k = l1-1; /* index for the sum */
    c = 0; /* Initial value of the carry */

/* addition is to be done from the right most digits */
    sum[l1] = '\0';
    while (i>=0){
        t = (n1[i] - '0') + (n2[i] - '0');
        sum[k--] = '0' + ( t + c) % 10;
        c = t / 10;
        i--;
    }
    return c;
}


/* Subtraction of BIG numbers */
int bigSUB(char n1[], char n2[], char sum[]) /* returns borrow */
{
    int l1, l2; /* length counter */
    int i, j, k, c, t, d;
```

```
    void insert(char[], int);
    l1 = strlen(n1); l2 = strlen(n2); /* get the length of the strings */
    d = l1 - l2; /* see the difference  and make them same */
    printf("%d %d %d\n", l1, l2, d);
    if ( d >0) { /* n1[] is bigger-- add d no. of 0's in front of n2[] */
        insert(n2, d); l2 = l2+d;
    }
    else if ( d < 0) { /* n2[] is bigger -- add d no. of 0's in front of n1[] */
        insert(n1, -d);
        l1 = l1-d;
    }

/* strings are adjusted; l1 = l2; now add digitwise from the right */
    printf(" %s\n", n1);
    printf(" %s\n", n2);
    i = l1-1;
    k = l1-1; /* index for the sum */
    c = 0; /* Initial value of the carry */

/* addition is to be done from the right most digits */
    sum[l1] = '\0';
    while (i>=0){
        t = (n1[i] - '0') - (n2[i] - '0');
        if (t>=0){
            sum[k--] = '0' + ( t - c) % 10;
            c = 0;
        }
        if (t < 0) {
            sum[k--] = '0' + ( t - c + 10) % 10;
            c = 1;
        }
        i--;
    }
        if ( c){
        i = l1-1;
        k = c;
        while( i > 0){
        j = 9 - (sum[i] -'0') + k;
        sum[i] = '0' + j%10;
                i++;
        }
        } */
    return c;
}

void insert(char s[], int n) /* support function for both ADD and SUBTRACT */
{
    int l;
    int i, j;
    l = strlen(s); j = l;
    for(i=1; i<=j+1; i++){
```

```
            s[l+n] = s[l];
            l--;
    }
    for(i=0; i<n; i++)
        s[i] = '0';
}
int main()    /* Test Code */
{
char  sum[101], n1[100], n2[100];
        int c;
        scanf("%s",n1);
        scanf("%s",n2);

    c = bigADD(n1, n2, sum);
        printf("%d%s\n", c,sum);

c = bigSUB(n1, n2, sum);
        printf("%d%s\n", c,sum);
return 0;
}


/*  Multiplication routine */
#include<stdio.h>
#include<string.h>
#define MAX 1000

int main(){
    char a[MAX];
    char b[MAX];
    char mul[MAX];
    int k;
    char *c;
    int la,lb;
    int i;
    int multiply(char [],char[], char[]);

    printf("Enter the first number : ");
    scanf("%s",a);
    printf("Enter the second number : ");
    scanf("%s",b);
    printf("Multiplication of two numbers : ");
    k =  multiply(a,b, mul);
    printf("%s",mul);
    return 0;
}


int multiply(char a[],char b[], char mul[]){
    char c[MAX];
    char temp[MAX];
    int la,lb;
    int i,j,k=0,x=0,y;
```

```
long int r=0;
long sum = 0;

la=strlen(a)-1;
lb=strlen(b)-1;

    for(i=0;i<=la;i++){
            a[i] = a[i] - '0';
    }

    for(i=0;i<=lb;i++){
            b[i] = b[i] - '0';
    }

for(i=lb;i>=0;i--){
    r=0;
    for(j=la;j>=0;j--){
        temp[k++] = (b[i]*a[j] + r)%10;
        r = (b[i]*a[j]+r)/10;
    }
    temp[k++] = r;
    x++;
    for(y = 0;y<x;y++){
        temp[k++] = 0;
    }
}

k=0;
r=0;
for(i=0;i<la+lb+2;i++){
    sum =0;
    y=0;
    for(j=1;j<=lb+1;j++){
        if(i <= la+j){
            sum = sum + temp[y+i];
        }
        y += j + la + 1;
    }
    c[k++] = (sum+r) %10;
    r = (sum+r)/10;
}
c[k] = r;
j=0;
for(i=k-1;i>=0;i--){
    mul[j++]=c[i] + '0';
}
mul[j]='\0';
return 0;
}
```

Bibliography

1. Bryant, R.E. and O'Hallaron, D.R., 2015. Computer System: A programmer's Perspective (3rd Edition). India. Pearson.

2. Courant, R., H. Robbins and I. Stewart., 1996. *What is Mathematics?* New York: Oxford University Press.

3. Gottfried, B., 1996. *Programming with C. Schaum's Outlines.* New York: McGraw-Hill Book Co.

4. Huntley, H.E., 1970. *The Divine Proportion.* New York: Dover Publications Inc.

5. Kernighan, B.W. and Ritchie, D.M., 2015. The C Programming Language (Second Edition). India. Pearson.

6. Livio, M., 2002. *The Golden Ration: The Story of PHI, The World's Most Astonishing Number.* New York: Broadway Books.

7. Mallik, A.K., 2018. *The Story of Numbers.* Bengaluru and Singapore: IISc Press and World Scientific Publishing Co. Pvt. Ltd.

8. Mallik, A.K., 2018. *Popular Problems and Puzzles in Mathematics.* Bengaluru: IISc Press.

9. Maor, E., 1999. *e: The Story of a Number.* Hyderabad: Universities Press.

10. Mano, M.M., 2015. *Digital Design* (5th Edition), India, Pearson.

11. Nisan, N. and Schocken, S., 2021. *The Elements of Computing Systems: Building a Modern Computer from First Principles.* Massachusetts: MIT Press.

12. Pickover, C.A., 2001. *Wonders of Numbers.* New York: Oxford University Press.

13. Roberts, J., 1992. *Lure of the Integers.* Washington D.C.: The Mathematical Association of America.

14. Sautoy, M.D., 2003. *The Music of the Primes.* New York: Harper Perennial.

15. Winkler, P., 2021. Mathematical Puzzles. Boca Raton, Florida: CRC Press.

16. https://projecteuler.net/

Index

Printed and bound by CPI Group (UK) Ltd, Croydon, CR0 4YY
01/05/2025
01858517-0001